ELEMENTS OF ELECT
BOOK

by
F.A. WILSON
C.G.I.A.,C.Eng.,F.I.E.E.,F.I.E.R.E.,F.B.I.M.

BERNARD BABANI (publishing) LTD
THE GRAMPIANS
SHEPHERDS BUSH ROAD
LONDON W6 7NF
ENGLAND

First Published – July 1981

British Library Cataloguing in Publication Data
Wilson, F.A.
Elements of electronics
Book 5, Communication
1. Telecommunications systems.
I. Title
621.38'0413 TK5101

ISBN 0 85934 064 3

Printed and Manufactured in Great Britain by Hunt Barnard Printing Ltd

PREFACE

Across the wires the electric message came . . .
Alfred Austin, 1835-1913.

Communication is the essence of life for without it we are nothing. For us therefore with an interest in electronics, it is an exciting subject to read or study. In this book we look at the electronic fundamentals over the whole of the communication scene; it is not the expert's book but neither is it for those looking for the easy way (there isn't one). The aim is to teach the important elements of each branch of the subject in a style as interesting and as practical as possible by using theory and mathematics where needed, yet always avoiding going too deeply.

As part of a series the text relies on the reader having a certain amount of basic electronic knowledge such as gained from Books 1 – 4 (especially 1 – 3) or as possessed by most up-to-date electronics people whether by hobby or profession. For those readers who already possess the earlier books, bracketed raised references in the form (Book/Section) are used as revision reminders. Written specifically for readers whose interest in electronics is awakening or those who need revision or updating, it is ideal not only for those who plan to enter a technical college or university and wish to do so with some basic knowledge but also for those who are unable to attend a college and are therefore obliged or prefer to study at home. In the series the level of mathematics is kept reasonably low but the subject is not avoided, it is taught as the reader progresses. The books all encourage the student to be practical, analytical and critical.

Switching, the complex process by which communication channels are connected together for people or machines to correspond, is not included because the transmission aspects alone fill the book. Nor do we concern ourselves with what the information is about, only how it gets there.

Brief descriptions of the earlier books in the series follow:

BOOK 1 **"The Simple Electronic Circuit and Components".** This deals with electricity, the electric circuit, electrostatics and electromagnetism backed up by several appendices teaching the required mathematics from arithmetic (for absolute beginners) through decimals to logarithms, simple mathematical equations and geometry.

BOOK 2 **"Alternating Current Theory".** Perhaps less mentally exhilarating but a must for all who wish to really get to grips with the subject. It considers the sine and more complex waveforms and how they react in the many basic alternating current circuits, also time constants together with the required mathematics, trigonometry and geometry.

BOOK 3 **"Semiconductor Technology"** prepares the way into the modern world of electronics with explanations of the working of semiconductors and of their practical characteristics, then of rectifiers, amplifiers, oscillators and switching (including a little on computers) with a chapter also on microminiature technology. The appendices cover the additional mathematics and binary arithmetic.

BOOK 4 **"Microprocessing Systems and Circuits"** branches out into the computer world. It is a book which really starts at the beginning of the whole subject and in this instance does not rely wholly on the preceding books, for non-electronically minded people there is a special appendix instead. A complete microcomputing system is explained bit by bit and both the need for and the execution of programming developed. This is followed by a comprehensive survey and explanation of the many basic electronic circuits in use. The appendices are written to provide an intimacy with numbering systems, especially binary.

CONTENTS

1. COMMUNICATION

Life is said to be nothing without music, it is even less without communication. Once we are outside the range of megaphones, smoke-signals and carrier pigeons and provided that we have no wish to suffer the delays of surface mail, then electronics must enter the scene. So we are confronted with an essential, interesting and ever expanding subject but with no pretence that it is an easy one. However if we avoid the more complex issues and mathematics while still involving ourselves to a degree sufficient to stretch our minds, we should emerge with an adequate understanding and with that extra sense of achievement.

For our purpose communication may be said to be the "transmission of information". The expert defines information as the "reduction of uncertainty" as though it takes something away rather than adding. The statement is meaningful however, for if there is no uncertainty, adding information cannot improve matters. For our part we will be satisfied with the more mundane idea of information being that which informs, tells or adds knowledge. Now to the very essence of our subject the *signal*.

Man's dependence on signals is as old as man himself, starting with his first cry as a baby to signal that life has begun. Our agenda however, restricts us mainly to the electronically generated signal which for the purpose of this book is defined as a voltage, current or electromagnetic form changing in some way to provide information, thus implying that without change a signal carries none.

One of the first attempts to assign numbers to something as abstract as information was published in 1948 by C.E.Shannon, an American mathematician. His ideas are generally known as *Information Theory*. It gives the communication engineer a means of estimating just how much information can flow over a particular path or *channel*. When a computer, for example, communicates with its peripheral equipment(4/1.3) this is normally over a short length (a few metres) of wiring. Little happens to signals travelling over such distances except that it takes them a few nanoseconds to do so. It is when greater distances arise that the transmission medium affects the signal in the ways we are about to discover so we start with the idea that channels vary in length from a few tens of metres to the many million kilometres of space exploration. The remainder of this chapter covers the main concepts superficially so that with a little prior understanding of the whole, we may more easily appreciate the functions of the parts when later we meet them.

1.1 SIGNAL TRANSMISSION

Signals are generated not only in all shapes and sizes but also with the characteristic of frequency. Fig 1.1 shows oscilloscope pictures of a range of typical signals. A pure sine-wave(2/1.3) has both constant amplitude and frequency and because it obeys a mathematical law it is easy to analyse, however as a bearer of information it has little use except from a knowledge simply of its presence or absence [Fig. 1.1(i)]. Our most common form of communication, the voice, carries more information because the frequencies generated by the vocal cords are not only changed in loudness or *amplitude* but are further modified by movement of tongue and lips. Clearly analysis or even measurement of such an irregular and uneven flow of information needs more sophisticated techniques than for the sine wave. However it is precisely because of these almost unquantifiable variations that such a signal carries more information. (We must remember here that sound itself does not travel, only pressure variations, these do not become "sound" until they have been sorted out by our ears).

Now for communication a talker needs a listener and there will always be some distance between them. One thing we quickly learn in life is that the greater this distance, the fainter the received speech becomes. In its passage through the air the signal power gets less, we say it suffers a *loss* or *attenuation* (Latin, to make thin). Also, especially when the speech is heard

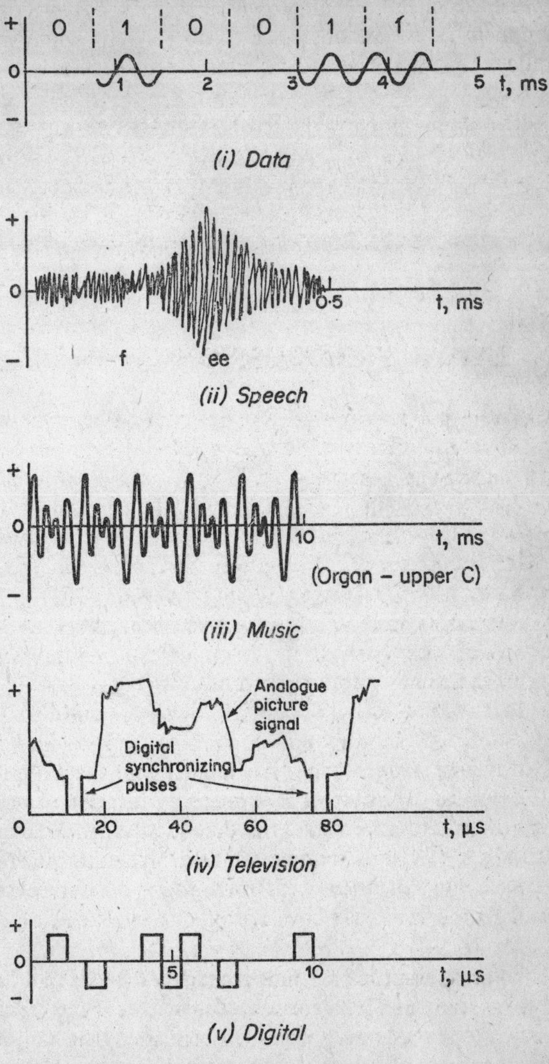

Fig. 1.1 Typical communication signals

faintly, noise can be a problem. As an example, two children may just manage to hear one another over a distance but will give up while a noisy aircraft flies overhead, the noise has "drowned" the signal.

These elementary ideas enable us to appreciate some of the characteristics of the *transmission channel* illustrated in Fig. 1.2. It forms a communication *link* between two people in conversation. The figure shows a *channel* as being effective in the direction A to B, for an air-path the B to A channel is the same one, in long-distance communication however this does not apply.

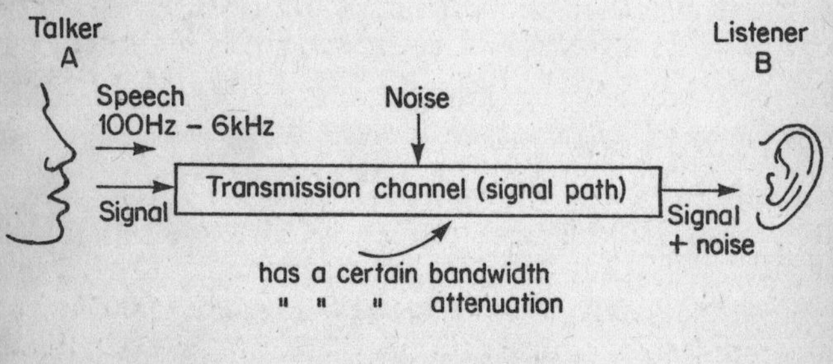

Fig. 1.2 Elements of a transmission channel

(i) speech frequencies lie approximately in the range 100 to 6000 Hz, thus the channel needs a frequency *bandwidth* of 6000 – 100 = 5900 Hz, meaning that for good communication all pressure variations, no matter at what frequency they occur within the range, must be reproduced.

(ii) the signals (the air-pressure variations representing speech) suffer attenuation in their passage through the channel.

(iii) noise may also be a factor because as in the example of the children, it can completely close down the channel or at least make conversation difficult. If so repetition is needed and effectively the rate of information flow has decreased.

We expand on these characteristics as this chapter unfolds so as to appreciate not only what a channel comprises but how it affects the signal itself. But so far we have only looked at speech as a signal, there are hosts of others, for example, all radio signals carry information. At present they extend upwards in frequency to over 200 gigahertz (200×10^9 Hz) with transmission channels confined mainly to the earth's atmosphere at the lower radio frequencies and to space at the higher. Computers have data transmitted to them or they may even send signals to each other over channels which are often telephone lines. This reminds us of telecommunication systems which carry many different types of signal, in fact all of those of Fig. 1.1 and over communication channels of just a pair of wires or a more complex arrangement such as a *carrier system* (radio frequencies over a cable or radio link), a *microwave link* (by radio at the higher frequencies), undersea cable, communication satellite or optical fibre – we have much to discuss.

1.2 BANDWIDTH

Speech apparently needs a bandwidth of about 6000 Hz for transmission over a channel without loss of "fidelity". Bandwidth costs money so the aim is to reduce the requirement as much as possible. Continuing with speech as an example, experience shows that a bandwidth of about half the quoted one is satisfactory for commercial use (e.g. telephony) where it is good enough for conversation without noticeable difficulty but certainly not "high-fidelity". Reducing the bandwidth much below 3000 Hz may cause two people conversing over such a channel some difficulty and even error. Similarly with a television picture, for this a satisfactory bandwidth is around 5 MHz, less produces a picture of inferior quality. Note the considerable difference in requirements for speech and a TV picture, we shall see later that this is because the information required to build up a picture far exceeds that for the reception of speech. Working on this same theme we should therefore expect music to require a greater bandwidth than speech because the sound of all the instruments of an orchestra must surely give rise to more information. Fourier analysis of a waveform shows that a square-wave signal such as a pulse (for example, representing a computer binary digit) theoretically requires an almost infinite bandwidth.(2/1.4.2) To provide this is simply not feasible, fortunately pulses are usually recognizable as such even when arriving somewhat disfigured, nevertheless bandwidths of high order are needed.

1.3 ATTENUATION

Signal transmission is effected mainly in one of two ways, by line (ie usually under ground or water) or by radio. A line has resistance so when a signal current flows through it there is a dissipation of power. The signal is therefore progressively reduced as it travels along the line, we say it is *attenuated*.

Consider a two-wire channel having wire resistances of say 300Ω working into a receiving circuit of 400Ω as in Fig. 1.3 (i) and let a signal of value 10V (frequency is unimportant) be applied at the sending-end. Then the signal current

$$i = \frac{10}{(300 + 300 + 400)} = 0.01\text{A}$$

In flowing through the receiving circuit this current produces a received signal voltage V_R of 0.01 x 400 = 4V, showing that the effect of the channel on the signal is to reduce its value by a factor of 0.4. Whatever signal voltage is chosen, this particular channel always has the same effect, a reduction of the signal voltage to 0.4 of its original value. Thus we can forget about signal levels and assess the channel by its voltage reduction factor (or gain if there is an amplifier somewhere). This appears to be a convenient method until we have several different channels in tandem when the various ratios have to be multiplied together. For example, if Channel A causes a reduction to 0.4 and Channel B to 0.25 as shown in Fig. 1.3 (ii), the overall signal reduction is 0.4 x 0.25 = 0.1. These are round figures chosen to simplify the example, seldom do such convenient values arise in practice.

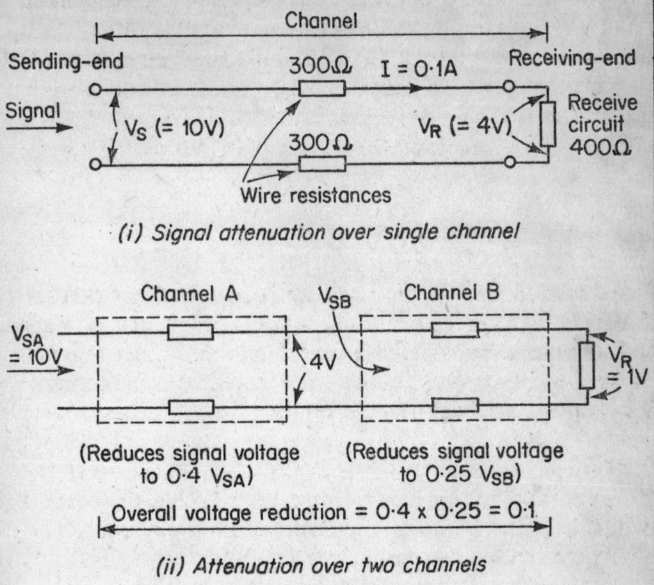

(i) Signal attenuation over single channel

(ii) Attenuation over two channels

Fig. 1.3 Signal attenuation

Logarithms are a convenient way of changing multiplication into the simpler process of addition, so how much easier it would be to work in logarithms of the ratios instead. The attenuations of Channels A and B could be quoted in the new logarithmic unit and simply added to calculate the overall attenuation. The unit is known as the *decibel*. The system is of such value in transmission engineering that we must understand it fully, it is considered in detail in Chapter 2.

The basic idea of attenuation continues to hold when radio signals are considered. Things get rather more complicated because we cannot measure the resistance of a channel which could, for example, be a few thousand miles of atmosphere. However, surprisingly enough, measurements can be made of the "strength" of the wave at both sending and receiving ends and the attenuation calculated.

For most channels the attenuation may vary over the frequency band, so for example, a pulse transmitted over a channel which has excessive attenuation at the higher frequencies will be distorted. The channel is said to possess *attenuation distortion*.

1.4 NOISE

The noisy aircraft mentioned in Section 1.1 provides an example of noise being injected into a transmission channel from outside. We all understand this as noise because it is objectionable and undesirable but not all electrical "noise" can be so described because sometimes it is generated for special purposes. But in general, noise is detrimental.

Noise in communication systems results from the advent of spurious (= not genuine) signals either injected from outside or generated within the channel. We call it a "signal" for once within a channel, noise receives the same treatment as does the wanted signal. We must clearly distinguish between audio noise and electrical, the latter is not necessarily audible although often it can be made so.(3/3.2.2.2)Even spots on a TV screen are "noise".

External noise is generated whenever a current changes rapidly as happens in relay circuits, automobiles, electric motors etc., and especially where sparking occurs. Fourier analysis shows that harmonics are generated, these enter a line channel through capacitive and/or inductive couplings. In radio communication some of the wide band of frequencies generated by a spark will be proper to a particular radio channel and so introduce noise.

Internal channel noise is generated thermally (white noise)(3/3.2.2.2) and by transistors, high-resistance relay contacts, joints etc. The total noise cannot be overlooked in channel design because an information signal must be significantly greater than an accompanying noise signal to avoid interference or at the limit being completely swamped. This point is developed in the next section.

1.5 INFORMATION THEORY

The formulae introduced in this section take into account the various features of a transmission channel which affect the rate of information flow, that is, they enable *channel capacity* to be calculated. The more information we try to squeeze through a channel, the greater the likelihood of error. We will not get involved in defining error because this complicates matters unduly, it may be easy to understand with computer signals but it is much more complicated with the analogue form such as speech.

Also the reader may not at this stage be fully at home with th. concept of bandwidth and it may seem that discussing information theory at the beginning rather than at the end of the book is putting the cart before the horse. It is presented in this sequence however because communication and information flow are inseparable and therefore to meet the basic principles of the latter first is of considerable help in developing a feeling for what everything later in the book is aiming at. As we progress, enlightenment on facets not fully understood here is sure to come.

1.5.1 Probability

Assume that we are at the receiving end of a channel and waiting for binary signals to arrive. At any point in time it is not known whether the next signal will be a binary 0 or a 1, for if it were known, there would be no need to transmit the information. The *chances* of it being either are assessed on a *probability* scale extending from 0 to 1 (nothing to do with the binary 0 and 1). There is a "fifty-fifty" chance of a binary 0 arriving so we assign to this probability a value of 0.5 and the same for a binary 1. This can be expressed as $P_0 = 0.5$ (the probability of a 0 arriving is 0.5) and $P_1 = 0.5$. $P_0 + P_1 = 0.5 + 0.5 = 1$. 1 in probability terms is certainty so we are sure that either a binary 0 or a 1 will arrive but only 50% sure that it will be a 0 or that it will be a 1. If there are four choices, say, of a 0, 1, 2 or 3 arriving (expressed by say, 4 different signal levels instead of 2) then the probability for any particularly one is 1 in 4, ie. 0.25. With a pack of 52 cards, the probability of any particular card being drawn from a properly shuffled pack is

$$1 \text{ in } 52 \text{ or } \frac{1}{52} = 0.0192 \text{ so we write}$$

$$P_{(10 \text{ of diamonds})} = 0.0192 \quad P_{(6 \text{ of spades})} = 0.0192 \text{ and so on.}$$

Thus, confronted with	choices	there is a probability of	that we could guess correctly what comes next or	that we would be wrong
	2	0.5	0.5	
	4	0.25	0.75	
	52	0.0192	0.981	

and evidently the probability of being wrong increases with the number of choices. Reasonably therefore, the more choices there are (lower P), the more information is given when the correct answer is transmitted.

1.5.2 Information Content

Take a single signal which can be either a 0 or a 1. From the above, $P_0 = 0.5$, $P_1 = 0.5$, and since the probability for all signals is the same we can write $P = 0.5$. The information content (I) of the signal is defined in the theory as

$$I = \log_2 P^{-1}, \text{ in this case } I = \log_2 \frac{1}{0.5} = \log_2 2 = 1$$

and this is called a *bit* (of information), the connexion with bits of computer data is evident. The formula shows that the information content is inversely proportional to the probability as would be expected from the foregoing section. What this means in simple language is that 1 bit of information is needed to distinguish between a 0 or 1, or, in general, between two *equiprobable* events.

The computer data is the simplest form, pulse or no pulse, how do we rate a more complicated signal such as for speech, music or television? Here we are faced with continuously varying signals and the information transmitted must simply state the level of the signal at any instant. The first decision required therefore is the number of *sampling* levels needed. This obviously must be greater than the two required in the binary case and yet must not be so large as to transmit unnecessary information, it costs money. A very simplified approach to this problem is at follows. The ear can only just detect a change in loudness equivalent to doubling or halving the signal power so there is little additional information gained from intervals less than this. (We must be careful here, doubling the signal power does *not* give the impression of "twice as loud", it needs much more than that). The range of speech powers from the loudest to that which can only just be heard is about 1000 : 1. Now 1024 is a doubling from 1 ten times ($2^{10} = 1024$) suggesting that the recognition of 10 different levels of signal power at any instant provides sufficient information. With 10 levels, the probability of any particular one occurring at any given time is 0.1, hence

$$\text{Information content, } I = \log_2 P^{-1} = \log_2 10$$

Now we meet the problem of logarithms to the base 2. Book 2(A4.2.2) shows how natural logarithms are obtained from common logarithms, the same method applies. i.e.

$$\log_2 x = \frac{\log_{10} x}{\log_{10} 2} = \frac{\log_{10} x}{0.3010}$$

$$\therefore \log_2 10 = \frac{\log_{10} 10}{0.3010} = \frac{1}{0.3010} = 3.32$$

$\therefore$ I = 3.32 bits (notice that compared with P = 0.5, I = 1, this lower probability gives more information).

For music in the form of an orchestra which may have a signal power ratio of up to 10^7, some 23 different levels at least need to be recognized, each sample therefore producing

$$\frac{\log_{10} 23}{0.3010} = 4.52 \text{ bits of information.}$$

Moving on to vision, the eye is found to be satisfied with about 8 shades of grey between black and white giving some 10 levels altogether, therefore the number of bits per black and white *picture element* is 3.32.

1.5.3 Information Flow

So far we are provided with a measure of the information in bits for one sample of the signal, the next question is obviously how frequently must these samples be taken? The theory gives a guide which is that at least one sample is required for each half-cycle of the signal waveform. To conform with this the *sampling rate* must therefore be at least twice the highest signal frequency. Thus for commercial speech with a maximum frequency of 3400 Hz, the minimum sampling frequency is 6800 Hz, giving an *information rate* of 3.32 bits at 6800 times per second, i.e. 22,576 bits/sec. (sampling at 8 kHz is used in practice, this is developed more in Section 5.5.3.3).

For high quality speech, 3.32 x 12,000 = 39,840 bits/sec.

and for music up to 10kHz, 4.52 x 20,000 = 90,400 bits/sec.

Countries have different TV standards, but taking the European one of 625 lines, the inference is that if the picture comprises 625 horizontal lines, each dot or picture element must be almost 1/625 of the height of the picture (for convenience we ignore the fact that not all lines are effective). Assuming that the width of a picture element is the same as its height, and since the screen width is 4/3 times the height, there are 625 x 4/3 elements per line. The total number of elements is therefore

$$625 \times 625 \times \frac{4}{3} = \text{say } 500{,}000$$

This is for a single picture, 25 complete pictures are built up in one second

∴ Information rate

$$= 3.32 \times 500{,}000 \times 25 = \text{say } 40 \times 10^6 \text{ bits/sec}$$

so a TV picture requires some 2000 times the information flow needed by commercial speech and we have not even allowed for colour. However, as shown below this is a maximum figure, generally a TV channel is equivalent to about 1000 speech channels, and at this cost a TV broadcasting company or authority does not have many of these spare!

Now we must be very clear that what we have just done is, as already mentioned, simplified. It merely leads us to ideas, based on Shannon's approach so as to get information rates for various transmitted signals into perspective. What is also lacking is any allowance for the fact that so many events are not equiprobable, for example with a television picture of an aeroplane in a cloudless sky, for many traverses of the spot showing the clear sky there is little uncertainty because each picture element is expected to be white. Moreover successive pictures may not change appreciably. Also with languages certain letters are much more likely to follow others (in English, for example, u is likely to follow q) and again, some are excluded (a z is most unlikely to follow a q), thus P does not equal 1/26 for all letters of the alphabet. With less uncertainty there is less information.

1.5.4 Channel Capacity.

The foregoing gives some indication of the rate of information flow for various types of signal. Channels invariably carry noise which, if of a sporadic form makes analysis difficult. However much can be done by assuming *white* or *thermal* noise(3/3.2.2) which arises from the continuous movement of electrical charges (also known as *Gaussian* noise because of its random nature – after J.C.F.Gauss, a German scientist). This makes itself heard as a hissing sound in the loudspeaker of a radio receiver when it is not tuned to a station and has the volume control at maximum, the receiver is then acting as a high-gain amplifier of the noise signal at the input.

The general formula from Shannon which enables us to focus on the main features affecting the transmission of information over a channel is

$$C = W \log_2 \left(1 + \frac{S}{N}\right) \text{ bits/sec (b/s)}$$

where

C is the channel capacity

W is the channel bandwidth

$\frac{S}{N}$ is the channel signal/noise ratio

and evidently it is not so much the noise level which matters but the degree by which the signal exceeds it. $\log_2 (1 + S/N)$ is given in Table 1.1 for a range of values of S/N for estimating the noise tolerable in a channel for a given rate of transmission without appreciable error. For example, for a commercial speech channel our estimate of C is 22,576 b/s. Therefore for a channel bandwidth of 3400 Hz:

$$22{,}576 = 3400 \log_2 \left(1 + \frac{S}{N}\right)$$

$$\therefore \log_2 \left(1 + \frac{S}{N}\right) = 6.64 \text{ and from Table 1.1, } \frac{S}{N} \approx 100.$$

which suggests that the mean value of the signal power should be at least 100 times greater than the mean value of the noise power. These are difficult quantities to measure considering the varying natures of the signals but it can be done.

It is also possible to see how the signal/noise ratio and bandwidth can be interchanged. Suppose a ratio of S/N no better than 10 is available. From Table 1.1,

$$\log_2 \left(1 + \frac{S}{N}\right) = 3.46$$

$$\therefore 22576 = W \times 3.46 \text{ and } W = 6525,$$

i.e. a bandwidth of about 6500 Hz will provide a sufficiently higher quality received speech signal to make up for the deterioration due to the increased noise.

Thus we observe mathematics getting to grips with the multitude of inconsistencies which seem to surround the subject of information flow. Information theory will seldom produce

TABLE 1.1 CALCULATION OF CHANNEL CAPACITY

S/N	$\log_2 (1 + \frac{S}{N})$	S/N	$\log_2 (1 + \frac{S}{N})$
0.1	0.138	1.5	1.32
0.25	0.322	2.0	1.58
0.5	0.585	3	2.00
0.75	0.807	5	2.58
1.0	1.0	10	3.46
		50	5.67
		100	6.66
		500	8.97
		1000	9.97

exact answers but we already see its usefulness in reminding us of the conditions necessary for optimum results. But we must not forget that we are only on the fringe of the theory and the examples have been designed for illustration only. So far though, we have discovered that:

(i) in its passage over a channel a signal is attenuated
(ii) if the attenuation varies with frequency the signal is also *distorted*. This also implies that the channel must have adequate bandwidth
(iii) noise is added to, or generated within a channel
(iv) channel capacity is affected by the signal/noise ratio and to a certain extent signal/noise ratio and bandwidth are interchangeable
(v) in addition, we already know that when a channel contains amplifiers, overloading may cause signal distortion.(2/1.4.3)

With so much signal deformation apparently possible, one may wonder whether signals received over long-distance channels are even recognizable. It is a tribute to modern communication engineering that so often quality leaves very little to be desired.

1.6 CABLES

Although we do not enquire into the technical aspects of lines until later in Chapter 6, it is worthwhile having a brief look at their construction first so that in discussion of land and sea links we can visualize the practical features better.

Before amplifiers were developed, long-distance circuits relied on heavy gauge overhead conductors to keep the circuit resistance and therefore attenuation low. It is difficult to appreciate now when we see the fine wires of today that overhead copper wires weighing in excess of half a tonne (or ton) per mile (about 0.3 tonne/km) lined some main roads. These wires had less than 1 ohm resistance per kilometre. Nowadays we tolerate high wire resistance and capacitance in the interest of economy, making up the signal attenuation by less expensive amplifiers. Both copper and aluminium wires are used, the latter is less expensive, is lighter but has a higher resistivity.(1/3.4.2) The external part of an underground or undersea cable is its sheath, previously of lead, expensive and heavy but a good electrical screen but now of some kind of plastic, usually polythene. Even two wires within a sheath constitute a *cable*. Wires are insulated from each other and the term "cable" usually implies that it is flexible.

1.6.1 Audio Frequency Cables

In "laying-up" wires together within a sheath for communication circuits (as opposed to power cables), capacitors are also formed when two wires run adjacent. The two-wire line is therefore equivalent to a resistance with capacitance in parallel, the very combination to cause an increase in attenuation with frequency. In addition two wires running parallel but belonging to different circuits also have capacitance between them and are a source of *crosstalk*, the term for the faint overhearing of conversations on other channels. However, provided that the capacitances balance so that capacitive currents from both wires of a circuit flow into the disturbed wire equally then being opposite, they cancel out Various means are employed to avoid any two wires running parallel over a long distance. In the *star-quad* type, [a typical cross-section of such a cable is shown in Fig. 1.4(i)] not only do the quads (that is four wires twisted around each other per unit) rotate but so do the layers. Diagonally opposite wires of a quad form a pair. In the *unit-twin* arrangement which to a certain extent is superseding the star-quad, as shown in Fig. 1.4(ii) the two wires of a pair are twisted together with different twist lengths for adjacent pairs. Groups of pairs (typically 50) are formed to make a unit with several units assembled within the sheath.

Small cables say up to 100 pairs may have thin polythene insulation on each wire with a polythene sheath overall Larger cables have paper lapping as insulation, it is important that this is kept dry otherwise insulation resistance falls. A polythene sheath itself is not completely impervious to water over a long period of time hence the use of a very thin aluminium foil moisture barrier on the sheath inside surface, this also restores

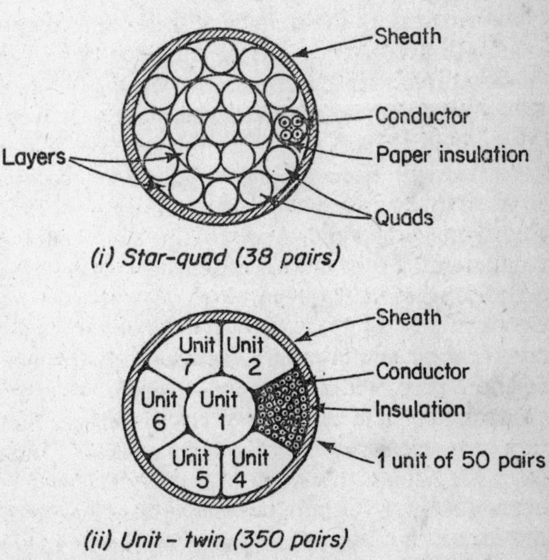

Fig. 1.4 Audio frequency cables

some of the electrical screening originally provided by lead. In addition cables may be pressurized with dry air, if a hole occurs anywhere in the sheath, air escaping prevents water entering, the drop in pressure is noted at the cable end and remedial measures put in hand.

Copper conductors used range between some 0.3 to 0.6mm diameter, aluminium are slightly larger. Cables may contain as many as 1000 pairs or more, special markings enabling the identification of each pair at both ends of a length. Smaller cables are sometimes buried directly in the ground but generally cables are pulled into pipes or *ducts*, the length in this case is limited (say up to about 400m, less for lead), otherwise pulling tension may tear the sheath. Thus manageable lengths are pulled in and all conductors jointed through, the joint then being housed in a *sleeve* which is made watertight on the sheaths of the two cables.

1.6.2 High Frequency Cables

As we progress up the frequency scale, various types of cable come into prominence, firstly pair-types which are specially manufactured with the conductor spacing increased so as to reduce capacitance, then *balanced and screened* pair-type cables in which each pair of polythene covered conductors is assembled with a second pair of polythene *wormings* (dummy wires) to make up a quad formation. Both wires and wormings are diagonally laid and an approximately circular cross-section is achieved. This is then lapped with a thin copper tape to provide an electrical screen.

Better known and in more widespread use both on land and at sea are *coaxial* cables. The main difficulty with the cables so far discussed is the *shielding* of any circuit against interference from other circuits within the same cable or from extraneous sources. An inner conductor surrounded by a concentric copper screen or "tube" (hence the term "co-axial") takes the place of a pair of wires, the tube itself is one of them and is always the earth or ground side of the circuit. To understand why a coaxial cable is more suitable at the higher frequencies we must look at both shielding and *skin effect* first.

1.6.2.1 Shielding

The need for shielding is evident when we consider the numerous neighbouring currents against which a circuit has to be protected. The earth itself contains a multiplicity, from power currents and their harmonics to all the communication currents we unintentionally put there, including radio transmissions. Although the latter are minute, an unprotected but insulated wire in the ground acts as an aerial and it so happens that many local radio stations broadcast on frequencies in the same range as used for some multiplex telephony systems. The voltages picked up may on average be a mere microvolt or less but near the radio transmitter voltages greatly in excess of this are found. As we progress in Chapters 4 and 5 we will begin to appreciate how damaging this could be.

The effect of a disturbing electric field(1/4.2.1) can be minimized by protecting other circuits with an earthed conducting *shield* or *screen*, usually in the form of a metallic sheath on the cable. The flux then terminates on the shield and the simplest way of looking at this is to consider that the flux is short-circuited by it. If magnetic flux(1/5.3) is also present the shield should have low magnetic reluctance(1/5.2.2) although at the higher frequencies this is not so important because eddy currents set up by the flux themselves produce a flux which mainly cancels the original one (Lenz's Law(1/5.3.1)).

1.6.2.2 Skin Effect

When a conductor carries a direct current there is no reason why the cross-sectional distribution should not be even. When the current is alternating however, the distribution changes. Consider the cross-sectional view of a conductor as shown in Fig. 1.5(i). At any instant in time the current flow causes a magnetic flux to be set up around the conductor(1/5.2.1) as shown by the dotted circles. As the current changes direction the flux first collapses, then is restored with the flux lines running in the opposite direction. We can well imagine the establishment of the flux as emanating from the centre of the conductor, growing outwards past its surface into the outside air, the opposite happening when the flux collapses. Thus the

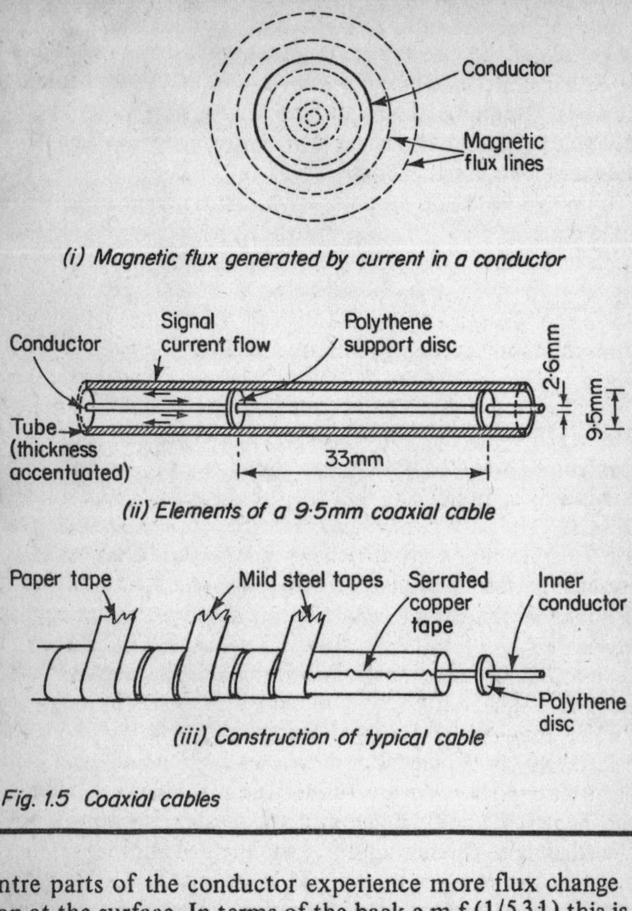

Fig. 1.5 Coaxial cables

centre parts of the conductor experience more flux change than at the surface. In terms of the back e.m.f.(1/5.3.1) this is greater at the centre than at the surface, hence the original current is forced to flow nearer the surface than through the middle of the conductor. Looked at in a slightly different way, the impedance of the conductor is greater at the centre therefore the magnitude of the current decreases towards the centre. Since inductive reactance, which is the greatest part of this impedance, increases with frequency, then skin effect also increases with frequency and the current path is forced more and more to the surface or "skin"of the conductor.

Obviously if the current cannot take full advantage of the whole cross-sectional area of a conductor, this is equivalent to a rise in resistance. The conductor then has an *effective* resistance R_{ac}. It can be shown that if R_{dc} is the normal resistance of a conductor (to d.c.) then

$$\frac{R_{ac}}{R_{dc}} \propto \sqrt{f}$$

1.6.2.3 The Coaxial Cable

From the two preceding sections we can now appreciate better Fig. 1.5(ii) which shows the basic features of one particular type of coaxial tube. The conductor of copper is held centrally within the copper tube by polythene discs so that in effect the dielectric is dry air. The flow of signal current along the centre conductor and along the tube, when at high frequencies, is forced by skin effect to be on the surfaces. Because the outer tube may in a way be looked at as wrapping the centre conductor completely within an earth connexion, the electric and magnetic fields generated by the signal terminate on the inside surface of the tube. The net result is that the tube current flows mainly on the inner surface. For the same reason external interference currents complete their journeys by flowing along the outside surface of the tube hence do not interfere with the signal current on the inner surface. Not quite the full explanation but sufficient for us to understand the benefit of the coaxial system, greater freedom from electrical interference and capacitance which impair the efficiency of pair-type cables. With the latter the difficulties increase with frequency, with the coaxial cable they decrease because of the greater separation of signal and interference currents.

One method of construction of a single tube is sketched in Fig.1.5(iii). The inner conductor is supported within the tube as already shown in (ii) of the figure. The tube itself is of copper and is formed longitudinally with a single seam running along its length, the serrations arranged so that the edges of the tape do not slip over each other. Directly over this are wrapped two mild steel tapes to add strength and increase the screening at the lower frequencies. Finally this is all lapped by a paper tape. Such coaxial tubes, up to 20 or more, are assembled in a cable along with groups of twisted pairs as required. Fig 1.5 shows the construction of a 9.5mm (inside diameter) tube, there are smaller sizes, one in common use being 4.4mm.

Undersea or *transoceanic* cables work on the same principles but differ in construction mainly in that the dielectric is solid polythene to withstand the very high undersea pressures. High tensile strength is also needed when the cable is being laid on the ocean bed from a ship on the surface, at depths of as much as 5km (over 3 miles). The strength is provided by using an inner conductor consisting of a bundle of steel wires lapped by a copper tape for conductivity. The outer sheath is polythene but with added steel wires surrounding it for near-shore lengths, this gives extra protection against tides, fishing trawlers and ships' anchors.

1.7 ELECTROACOUSTIC TRANSDUCERS

Although as time goes by communication systems work more and more to electronic devices for data, computer, television, teletype, etc., very much is still with us, the non-electronic humans. Our main communication senses are speech, hearing and sight, the first two predominate thus it is befitting that we should look briefly firstly at our own wonderful communication endowments, then at the types of electroacoustic transducers. These put the fleeting sound waves into wires for transmission afar and then give them back again to us.

1.7.1 Human Communication

It all starts in the *larynx*, the cavity in the throat which holds the *vocal cords.* The *trachea* or "windpipe" travels up through the throat from lungs to the nose and mouth. When air is being exhaled it passes through the vocal cords, a pair of membranes fixed to the walls of the pipe rather like a diaphragm with a slit, somewhat less than 2cm long, from front to back. The membrane tissue is elastic and the walls of the slit vibrate when air is forced upwards between them. Muscles control the tension in the cords while sound is being voiced and the pitch of the sound is controlled by the tightness of the cords. The loudness produced depends on how hard we force the air through the cords, we can in fact sense the extra effort required when we shout.

The sound itself is simply air travelling upwards from the vocal cords as a rapid train of pulses, about 100 – 200 per second for a male voice and some 150 – 300 per second for a female. This forms the fundamental frequency. Many harmonics are also present and while voicing sound these air pulses are reinforced and modified considerably by resonances in the vocal tract above the cords and by the positions of tongue, lips and teeth to give the multiplicity of sounds which we are capable of uttering.

The sound waves are airborne, air being an elastic medium which is alternately compressed and rarefied at a rate according to the wave frequency. Progress of a wave is indicated at a point by the *sound pressure* which is a measure of the degree of fluctuation above and below the ambient atmospheric pressure. Sound waves travel at about 344m/s at 20°C (temperature affects the velocity slightly), very slowly indeed compared with radio waves.

Hearing must firstly depend on sensing the minute variations in air pressure, this is done by a tiny diaphragm a little less than one square cm in area and known to all as the *eardrum*. The alternate compressions and rarefactions of the air move the drum in sympathy, so it vibrates in frequency and amplitude according to the incoming sound wave. Coupled to the drum inside the ear is a tiny chain of bones along which the vibrations are carried to the inner ear. This contains the complex mechanism which separates the various frequencies by stimulating different nerves, these then transmit the frequencies separately to the brain in the form of minute electrical impulses. When they reach the brain, we hear *sound.*

1.7.2 Microphones

A *microphone* is a transducer for the conversion of sound waves into the corresponding electrical signals. It is not unlike the ear in its method of sensing the wave, it too has a diaphragm which, like the eardrum, vibrates when a wave impinges upon it. The differences between the various types of microphone discussed in this section are in the process by which the diaphragm movement is converted into electron flow. Fig 1.6 shows in a much simplified pictorial fashion the several main types. Except for the first one, the voltage output of a microphone for average signal input is in the range of millivolts or just below and frequency response is from some 50 Hz or less up to 15 kHz or more.

Figure 1.6

(i) *carbon-granule*: is a variable resistance or *loose-contact* microphone, very much in favour for telephone instruments. Loosely packed between two carbonised nickel electrodes are small carbon granules with a resistivity in the semiconductor range and obtained by processing certain hard coals. Movement of the diaphragm by the action of a sound wave varies the pressure on the granules, packing them more tightly or more loosely and hence varying the resistance of the complete granule chamber so for a fairly constant current I, producing a varying voltage E as shown. Some 20 – 100 mA direct current flows through the chamber for satisfactory output.

Carbon-granule microphones have the distinct advantage of high output (some 1V can be obtained compared with millivolts for most other types) but the disadvantages of noise (current flows through thousands of tiny contacts), instability, poor frequency response (about 200 Hz to 4 kHz) and harmonic distortion.

(ii) the *moving-coil* microphone needs no battery supply because it is a generator. A plastic or aluminium foil diaphragm has a coil former attached to it at the centre as shown which is free to move between the pole-pieces of a circular permanent magnet. The magnetic flux is at right-

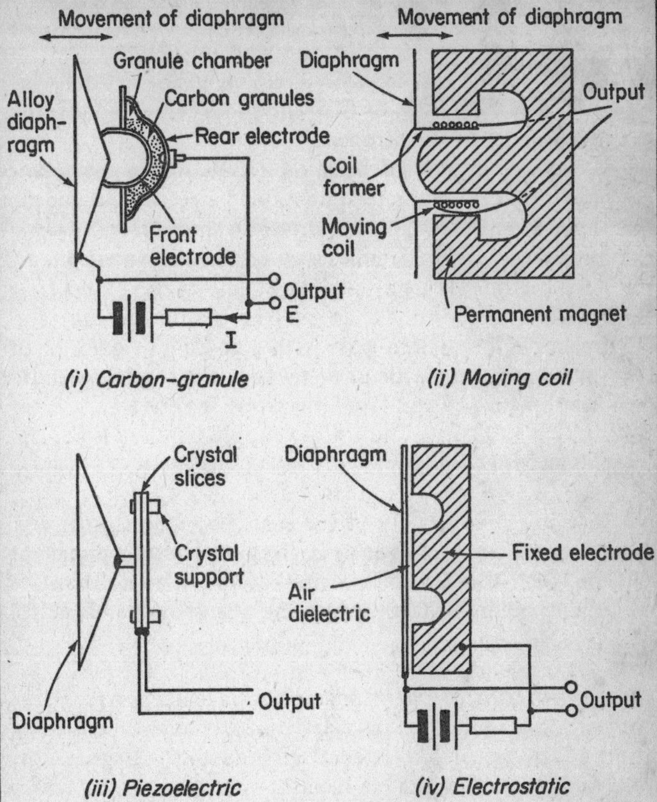

Fig. 1.6 Microphones

angles to the turns of a coil of fine wire wound on the former, it is in this coil that the e.m.f. is generated when the diaphgram vibrates.(1/5.3) Because the voltage produced depends on the *rate* of cutting lines of force, hence to the velocity of the wires, this is termed a *velocity* microphone.

(iii) *piezoelectric* microphones depend on the effect of

mechanical stress on certain crystals, for example, Rochelle salt or specially developed polycrystalline ceramics. When stress is applied in a certain manner, the crystal develops electrical charges.(3/3.3.3) The diaphragm is mechanically coupled to the crystal which may be in the form of two thin slices cemented together with metal foil electrodes.

(iv) *electrostatic* types depend on a variation in capacitance between the diaphragm and a fixed electrode. A constant polarizing potential is required and since $Q = CV$,(1/4.1.1) any variation in C implies a corresponding variation in Q, the quantity of electricity stored and current must flow when this quantity changes. The diaphragm may for example, be of thin glass with a coating of gold, or of metal-sprayed plastic or metal foil. The gap between the diaphragm and the fixed electrode is very small, some 0.02 – 0.03 mm. The voltage supply required is at least 50 and for some types very much more.

Electret microphones are of the same basic type but use a plastic film diaphragm which is given an almost permanent charge when the device is manufactured. The constant polarizing potential to maintain the charge is therefore not required.

Electrets or other types of microphone are likely to replace the carbon-granule type in telephone instruments because the great advantage of the latter of high output voltage is now nullified by the fact that an electret microphone for example, plus a high-gain transistor amplifier can do the job even better. Amplifiers are now small enough to be hidden within the microphone case and at low cost, hence the greater stability, frequency response and freedom from noise of the electret and others make such a change attractive.

1.7.3 Earphones

An *earphone* is an electroacoustic transducer designed for conversion of an electrical signal into its corresponding sound waves and for use directly on the ear. A single earphone may be used as in a telephone handset or a pair used together held by a headband. In the types we look at the principle is simply the setting up of sound waves by causing vibration of a diaphragm, the reverse of the microphone principle. Except for the first which is designed for telephones, earphones are effective over a frequency range from as low as 20 Hz to over 20 kHz, more than sufficient for the average listener for it is a young and sensitive ear indeed which can benefit much from a sound wave at over 20 kHz.

Fig 1.7 (i) shows the elements of the *rocking-armature*

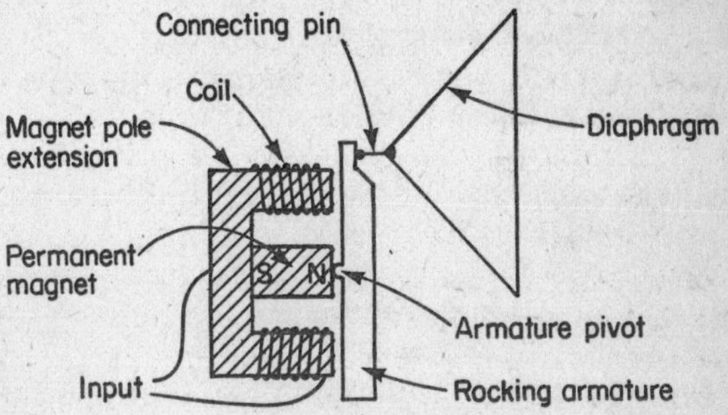

(i) Rocking-armature

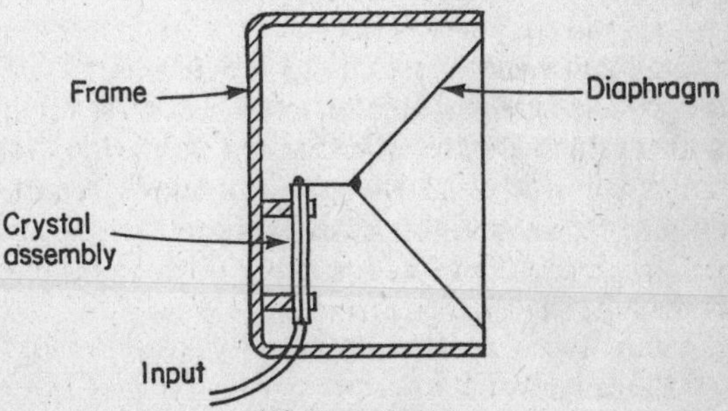

(ii) Piezoelectric

Fig. 1.7 Earphones

telephone receiver. The diaphragm is of a light non-magnetic alloy and is connected by a pin to one end of the armature. The latter is held against one pole of a permanent magnet and spaced from it by a pivot on which it can rock. The coils are oppositely wound and connected in series so that they exert a push-pull action on the armature. This communicates its movement via the connecting pin to the diaphragm, vibration of which sets up sound waves. This type of receiver has a frequency response effective up to about 4 kHz.

Moving-coil – this relies on the interaction between the magnetic field set up by the signal current in a coil and that of a permanent magnet, the motor principle.(1/5.6) The elements of such an earphone are as shown for a moving-coil microphone in Fig. 1.6(ii) with the exception that the output wires become the input. The physical construction is naturally different, for example, the diaphragm of a microphone must be light and unfettered for the pressure which moves it is extremely small, the diaphragm of an earphone is not so restricted because the power driving it is much greater, it is therefore of more robust construction.

A *piezoelectric* receiver is shown in Fig. 1.7(ii), the crystal slices distort when a voltage is applied and move the diaphragm. This type is particularly useful when a high impedance is required.

In the *electrostatic* types [see Fig. 1.6(iv)] a polarizing voltage is applied between the diaphragm and the fixed electrode. Alternating signals added to this voltage result in a changing force on the diaphragm which therefore moves in sympathy. The *electret* principle is also applicable.

1.7.4 Loudspeakers

We have in fact now covered in the above section most of the basic methods on which loudspeakers also operate. Generally they are of the moving-coil type as illustrated in Fig. 1.8. The diaphragm or *cone* is frequently constructed of stiff paper, not necessarily circular, elliptical units are often employed in small radio receivers. Attached to the diaphragm is the

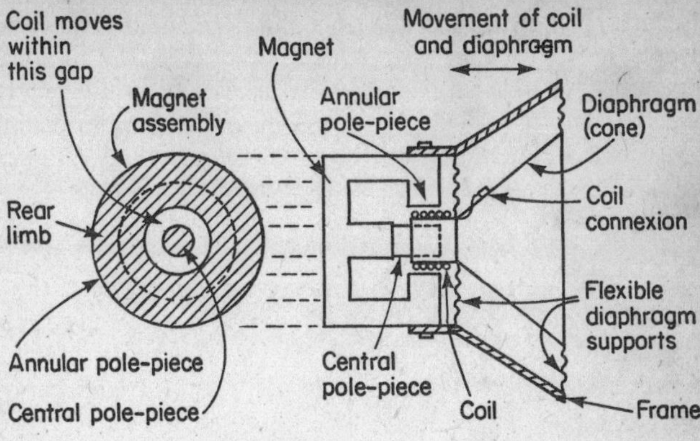

Fig. 1.8 Moving-coil loudspeaker

moving coil within the strong radial magnetic field. The magnet is usually circular with a central pole-piece projecting within the coil-former. The units are available with very small cone sizes (a few cms), at this size resembling a moving-coil microphone (a single unit is sometimes switched to serve either as a microphone or loudspeaker) up to 50cms or more. It is difficult to design a single unit to cover the whole audio range, the technique usually being to use a large cone speaker to handle the lower frequencies and a smaller size for the higher with a *cross-over* unit (filter) to direct each audio band to the appropriate speaker.

Other principles are used such as electrostatic and piezo-electric, generally these are specialized units.

2. TRANSMISSION QUALITY ASSESSMENT

Because communication systems serve people it is only they who can truly assess the performance and value of a system. But we as users not only have different requirements of a channel but our own faculties and mannerisms such as loudness of talking, acuity of hearing and the way we hold a telephone handset are also so diverse that assessment of the quality of a radio link, telephone connexion or even a t.v. picture needs many people in the tests so as to get an "average" result.

Furthermore what unit of measurement can we use? We have absolute units for so many things such as length, weight, heat and light but we have yet to discover one for the quality of a communication channel. So, generally the technique is to rate the channel under scrutiny against a reasonably similar standard one but again the question arises as to units. Simply to say something is twice as good as something else is of no value whatsoever because we do not know what "good" means. The many problems are evident, we can only touch upon them briefly in this chapter. However we shall find that even when human beings assess the quality of a system the decibel is often firmly implanted, therefore we discuss this unit in greater depth.

2.1 TRANSMISSION MEASUREMENTS

In Chapter 1 the idea of the decibel is introduced and as might be expected, there are few problems when continuous sine waves are concerned but many with some of the more variable signals. Channels are usually rated *objectively* (i.e. by measuring instrument as opposed to *subjectively* when human observers are used) according to their treatment of a sine wave test signal because of the convenience of calculable, repeatable results. In this section we get to grips with *decibel notation* which is exclusively a transmission unit and therefore very relevant to our later studies.

We talked in terms of signal voltage in Section 1.3 but now consider a signal appearing at the output terminals of an amplifier or at the end of a pair of wires in a cable and suppose that looking back into the amplifier or cable the resistance is 1000Ω and the signal potential measures 1 volt. We do this again for a second signal where the circuit resistance is 100Ω and the signal again measures 1 volt. We might be tempted to class the two signals as being of equal strength because they both exhibit the same voltage but some doubt would creep in on the realization that signal 2 is providing a heavier current than signal 1 because it is operating in a lower resistance.

If signal 1 maintains 1V across a 1000Ω circuit, the power being dissipated (V^2/R)) is 1mW and we call this the *signal power*. If signal 2 maintains 1V across a 100Ω circuit the power is 10mW so on a strength or power basis signal 2 is greater. Let us confirm this by putting signal 2 into signal 1's circuit, that is, a power of 10mW in a circuit of 1000Ω:

$$P = \frac{V^2}{R}$$

$$\therefore \quad V = \sqrt{RP} = \sqrt{1000 \times 0.01} = \sqrt{10} = 3.16 \text{ volts,}$$

so under the same conditions of 1000Ω, signal 2 exhibits a higher voltage. The moral of this should be clear, it is that signals ought really to be compared on a power basis but we can use voltage (or current) provided that we bring the circuit resistance or impedance into the picture.

2.1.1 Transmission Units

Section 1.3 suggests that a logarithmic transmission unit has operational advantages and from the above, sensing that we ought at least to start on the basis of power, then for any system the transmission unit is simply the logarithm of the ratio of the powers at two different points in a circuit or channel. The unit is called the *bel* (after Alexander Graham Bell, the Scottish inventor) and

$$\text{No of Bels} = \log_{10} \frac{P_2}{P_1}.$$

The unit happens to be rather large so one-tenth of this is use instead, the *decibel* (dB)

$$\text{No of decibels} = 10 \log_{10}\frac{P_2}{P_1}$$

(for $\frac{P_2}{P_1}$ we can also write

$$\frac{\text{Power Out}}{\text{Power In}}, \frac{\text{Power Received}}{\text{Power Sent}} \text{ etc.)}$$

A little revision on logarithms(1/A3) may be useful first. Common logarithms are to the base 10. The logarithm of 1 is 0, of 10 is 1, of 100 is 2, of 1000 is 3. Logarithms of numbers are given in logarithm tables, thus, for example, the logarithm of 586 which is between 100 and 1000 must be between 2 and 3. Logarithm tables show the value 0.7679 which must be added to 2 to produce the anwer between 2 and 3. Log. 586 is therefore

$$2.7679 \text{ (i.e. } 10^{2.7679} = 586)$$

Returning to the formula for decibels (it is one well worth committing to memory), if a 1mW signal at the input of an amplifier gives an output of 80mW then

$$\text{Amplifier gain in decibels} = 10 \log\frac{80}{1} = 10 \times 1.9031$$

$$= 19.031, \text{ i.e. approximately } 19\text{dB}.$$

To find one of the powers if the other and the gain (or loss) in decibels is known, say $P_2 = 0.5\text{W}$ and the circuit *loss* is 27dB then

$$27 = 10 \log \frac{0.5}{P_1}.$$

Here we take the antilogarithm of both sides because the antilogarithm of a logarithm re-establishes the original number, thus

$$\text{antilog } 2.7 = \text{antilog } (\log \frac{0.5}{P_1}) = \frac{0.5}{P_1}$$

$$\therefore\ P_1 = \frac{0.5}{\text{antilog } 2.7}\ \text{W.}$$

Antilog 2.7 must be somewhere between 10^2 and 10^3, from tables, antilog 0.7 = 5012, therefore antilog 2.7 = 501.2.

$$\therefore\ P_1 = \frac{0.5}{501.2} \times 1000\,\text{mW} = 1\text{mW.}$$

meaning that a signal power of 500mW experiencing a loss of 27dB is reduced to 1mW. Equally a signal power of 1W is reduced to 2mW, i.e. 27dB represents a power ratio of about 500. This is important, it does not matter what the actual powers are, decibels only express the ratio between them. Another advantage of decibel notation is now apparent, large power ratios are converted in decibels to much smaller and convenient numbers.

The usefulness of the system is demonstrated by a simple exercise. Suppose a cable circuit has a loss of 12dB and it is connected to an amplifier with a gain of 18dB. By using a negative sign for loss and positive for gain, the overall loss or gain for the cable plus amplifier becomes $-12 + 18 = +6$dB, i.e. a gain of 6dB and how a 1mW signal applied to the cable would appear at the amplifier output is calculated from

$$6 = 10 \log \frac{P_{out}}{1}$$

$$\therefore\ P_{out} = \text{antilog } 0.6 = 3.981\text{mW.}$$

In the above loss has been given a negative sign. This comes automatically from the formula for if, for example, a 1mW signal power input results in 0.5mW output,

$$\frac{\text{Power Out}}{\text{Power In}} = \frac{0.5}{1} = 0.5 \text{ and}$$

$$\text{No of decibels} = 10 \log 0.5 = 10\,(\bar{1}.6990)$$

(remember the characteristic is negative but the mantissa positive – these manipulations are fully covered in 1/A3)

$\therefore$ no of dB = $10(-1 + 0.6990)$ = $10(-0.3010)$ = -3.01.

This is doing the job properly. But a short cut which avoids juggling with characteristics and mantissas is to remember that loss is negative, gain is positive and always arrange the ratio of the two powers to be greater than 1. Thus for 1mW input and 0.5mW output we use the ratio

$$\frac{1}{0.5} = 2 \text{ and}$$

$$\text{no of dB} = 10 \log 2 = 10 \times 0.3010 = 3.01$$

and because we knew all along that it must be a loss we precede the answer by a minus sign, i.e. –3.01dB. The output power is then said to be at –3.01dB relative to the input power or alternatively that there is a 3.01dB loss.

Some power ratios and their equivalents in decibels are

Power Ratio	0.001	0.01	0.1	0.25	0.5	1	1.25	2	4	10	100	1000
dB	–30	–20	–10	–6*	–3*	0	1*	3*	6*	10	20	30

*slightly inaccurate but useful as a guide.

2.1.1.1 Voltage and Current Ratios

Decibel notation does work with voltage or current ratios provided that either the input and output resistances are the same or the difference is taken into account for

$$\text{No. of decibels} = 10 \log \frac{P_2}{P_1} = 10 \log \frac{I_2^2 R_2}{I_1^2 R_1}$$

$$= 20 \log \frac{I_2}{I_1} + 10 \log \frac{R_2}{R_1}$$

(remember $\log x^2 = 2 \log x$)

$$\text{Similarly, no. of decibels} = 20 \log \frac{V_2}{V_1} + 10 \log \frac{R_1}{R_2}$$

Thus when $R_1 = R_2$ (remembering that $\log 1 = 0$),

$$\text{No. of decibels} = 20 \log \frac{I_2}{I_1} \text{ or } 20 \log \frac{V_2}{V_1}$$

Voltage gains of amplifiers are often expressed in this way. Line amplifiers are usually designed to have equal input and output impedances but many others do not have such equivalence. For example, a power amplifier driving a loudspeaker has a moderately high input but low output impedance. To simply quote the decibel gain from

$$20\log_{10} \frac{V_{out}}{V_{in}}$$

is therefore incorrect although often done, when this is done it is important to state the fact that different impedances are involved.

2 1.1.2 Reference Levels

Absolute values can be quoted in decibel notation provided that a *reference* or *zero* level is stated or known. The reference level of some quantity such as signal power or sound pressure is chosen and this is given the decibel value of 0. The reference level is sometimes indicated by an added letter for example, using a reference level of 1mW for transmission measurements, other levels are quoted in *dBm*. Thus a power level of 100mW is correctly expressed as +20dBm, for a power gain of 20dB on 1mW is 100mW. No reference level need be quoted because it is indicated by the "m". Equally –30dBm is the same as 1μW.

A transmission unit also frequently encountered is the *Volume Unit* (vu), it is usually associated with speech signal measurements. Because speech has a complex waveform comprising many different and varying frequencies and amplitudes it is measured by using a special meter in that it has a certain delayed pointer rise-time (it rises to 99% of final reading in 300ms) to iron out minor fluctuations. This helps trained observers when watching the pointer to estimate a vu value for a given sample of speech. The reference level is again 1mW (0 on the scale) and levels in dB above this are marked as volume units, The vu is defined by

$$\text{No. of vu} = 10\log\frac{P_2}{0.001}$$

when P_2 is expressed in watts, or $10 \log P_2$ when P_2 is expressed in mW.

2.1.1.3 Nepers

Used in some countries and especially in theoretical work the *Neper* (after John Napier, a Scottish mathematician) is a transmission unit rather similar to the decibel except that natural logarithms are used with the ratio of two currents (or voltages) but with the impedances in which the currents flow not taken into account.

$$\text{No of nepers} = \log_e\frac{I_2}{I_1} \quad \text{and no. of decinepers} = 10 \log_e\frac{I_2}{I_1}$$

Tables of natural logarithms (to the base e = 2.71828) are available for calculations, alternatively common logarithm tables may be used since the natural logarithm of any number x is given by

$$\frac{\log_{10} x}{\log_{10} e} = \frac{\log_{10} x}{0.4343} \text{ or } \log_{10} x \times 2.3026 \quad (2/A4.2.2)$$

We obtain a relationship between nepers and decibels as follows. For n dB:

$$n = 20\log_{10}\frac{I_2}{I_1}$$

and by changing from common to natural logarithms

$$n = 20\log_e\frac{I_2}{I_1} \times \log_{10} e = 20\log_e\frac{I_2}{I_1} \times 0.4343$$

$$= \log_e\frac{I_2}{I_1} \times 8.686 = \text{no of nepers} \times 8.686$$

from which 1 neper is equivalent to 8.686dB – but this comparison is only valid when the measurements are made in the same input and output impedances.

We shall get greater experience and a feeling for transmission gains and losses as we progress.

2.1.2 Assessment of Channel Performance

A manufacturer who sells his products to the public usually has an automatic assessment of their value, if high, they sell, if low, they do not. Communication channels however are not sold in this way, unless the channel is faulty the user has no option but to accept and use it. Early telephones over which people had to shout were then just as acceptable as are those of today. Thus the design engineer in striving for the best service consistent with cost must have some means of knowing how satisfactory the product is, this is what the title of this section means. As a radio or line channel the product cannot be held in the hand and examined but fortunately fairly representative laboratory models of even world-wide channels can be built and tested.

2.1.2.1 Measurement of Transmission Gain and Loss

The basic technique of measurement in the field is invariably by using sine waves (also called *pure tones*) because of the comparative ease with which this can be done. On the other hand, laboratory assessments are closer to real life, generally involving both expert and untrained participants, to talk and listen over channels, view television etc. Their results have then in some way to be related to pure-tone tests in the field.

With digital transmission the techniques differ for the ultimate criterion is usually error, meaning that a received digit is wrongly interpreted, a serious condition with computer data but as we shall find later, less serious for digital systems carrying speech.

Returning to analogue signals, any channel which transmits them can be considered as a 4-terminal black box having a gain or loss which varies with frequency. Fig. 2.1(i) shows such a representation of a channel. Sine waves applied over a range of frequencies but at constant level result in an output/frequency characteristic as shown for a typical cable circuit.

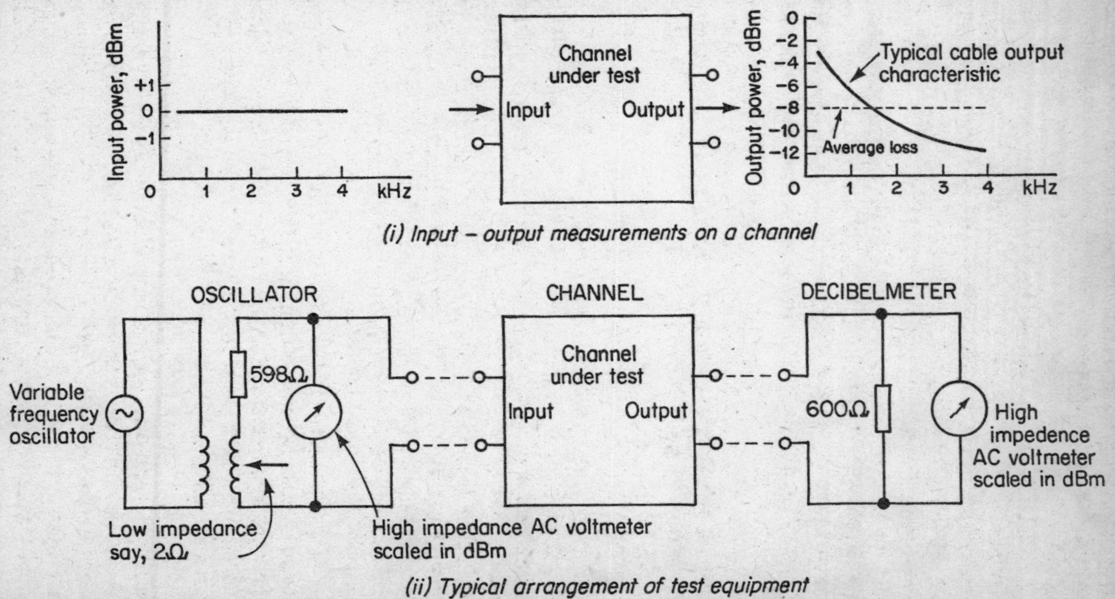

(i) Input – output measurements on a channel

(ii) Typical arrangement of test equipment

Fig. 2.1 Measuring overall channel gain or loss

What perhaps is of first concern to a user of a speech channel is the overall loudness. This is closely allied to attenuation, an examination of which therefore leads to a fair estimation of the channel performance. Due regard must be taken as to any variation with frequency, for example the channel in Fig.2.1(i) might be estimated to have an 8dB loss averaged over the speech frequency range.

For a channel not to degrade overall performance, a loss of 0dB at all channel frequencies is clearly the aim (output equal to input), failing this the minimum which can be achieved, thus pure-tone measurements indicate how well the objective is met.

A test-signal generator is a sine wave oscillator with its frequency variable continuously or in steps over the required range and with some means such as a built-in meter for ensuring that the test signal power applied to the channel is at constant level, usually 0dBm. The level should not be so low as to be affected by noise nor high enough to overload channel amplifiers hence the choice of 1mW which falls conveniently between. Standard sending and receiving resistances are usually employed as an average to suit all channels, such as 600 or 900Ω. At the output of the channel a *decibelmeter* is used which closes the circuit with the standard resistance and indicates the test signal power in the resistance by measuring the voltage developed across it. Fig. 2.1(ii) shows a typical arrangement. In the figure at 0dBm on the scale representing 1mW of power in 600Ω, the a.c. voltmeter shown actually measures $\sqrt{RP}$ volts, i.e. $\sqrt{600 \times 0.001} = 0.7746$V. Measurements are made at suitable frequencies in the range and a characteristic such as shown in (i) of the figure obtained. The oscillator and decibelmeter may be a great distance apart thus the testing staff need a separate telephone link between them. For quick checks a single frequency of say, 800 or 1000Hz may be used. Practical channel losses are considered in Chapter 4.

This is just one representative type of measurement, from what has been said already, there must also be methods of measuring noise picked up or generated by a channel. For digital systems it is also necessary to see how well a pulse travels, thus there is a whole range of test equipment, such as noise generators, measuring sets, chart recorders, pulse generators and oscilloscopes to name only a few.

2.1.2.2 Conversation Tests

Telephone links are never quite up to the standard of a direct air path (say, for two people one metre apart) because even with a low loss connexion, more repeats and spelling-out tend to be requested. This would be expected from the single fact that transducers limit frequency response. Things get worse as circuit loss increases and although human beings are quite tolerant of communication links and most will manage over a moderately bad connexion, there comes a point where some complain. There is however, no clear dividing line between what people consider as good enough and what is unacceptable. Thus assessment by human beings is necessary because although pure-tone tests can reasonably predict overall loudness, they cannot tell us how satisfactory a telephone or radio link is when degradations are in the form of noise, distortion, crosstalk or delay. We begin to suspect that assessment of communication links is almost as much an art as a science. As such therefore, we simply select one technique which is proving of value just to get an idea of some of the background work which goes on in the communications laboratory.

Such tests are necessary to build the basis for planning a telephone network. "Standard" links are chosen involving the current telephone instrument with circuit and/or surrounding (room) noise, distortions, delays etc. added artificially. Untrained subjects, acting purely as telephone users and two at a time are accommodated each in a small room or cabinet, insulated from extraneous noise. The two cabinets are connected together via the test channel. Each cabinet contains the test telephone instrument, a loudspeaker for introducing room noise if required, some separate form of telephone link with the experiment observer and material to help generate conversation. The latter is required because two people (who may be strangers), when asked to converse with each other tend to "dry up". To encourage conversation several methods have been developed often involving the solution of simple puzzles. Solving the puzzle is of no importance, talking over the connexion is. At the conclusion of each test call the participants are asked whether they had experienced any difficulty in the conversation, a simple yes/no answer required. From the results from several pairs of subjects the percentage experiencing difficulty is calculated. This type of test can be repeated for different channels, telephone instruments and channel attenuations thus comparison between tests shows the effect of some change. As a single example, curves for variations in channel attenuation with different levels of injected channel noise are given in Fig. 2.2. The absolute levels of attenuation and noise are unimportant here, we need only be concerned with the differences. The curves are the result probably of many weeks of testing followed by statistical analysis. They do not tell a designer what value of P (percentage of users experiencing difficulty)) to work to, the aim can only be to reduce it as the communication network develops, and with due regard to cost. Certainly to bring P

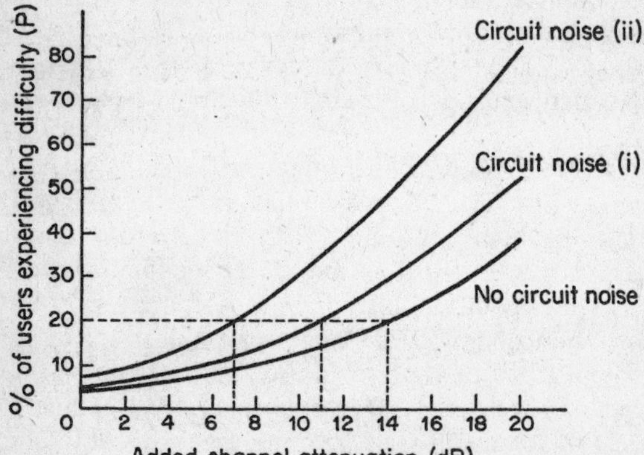

Fig. 2.2 Effect of circuit noise and attenuation on percentage difficulty

down to 0 would be impossible. What these particular curves do indicate is the equivalence between channel noise and attenuation, thus taking for example, P = 20%, some 14dB channel attenuation is possible with no noise present. For circuit noise (i) it is only about 11dB and for circuit noise (ii) which is some 10dB higher than (i), about 7dB (these points are shown dotted in the figure). This therefore rates the extra loudness people need to overcome the various levels of noise.

As a practical example, considering a trunk cable having an attenuation of 0.7dB/km, then the maximum length tolerable of such a cable if subject to pick-up of noise at circuit noise (ii) level is

$$\frac{14-7}{0.7} = \text{10km less than if no noise were present. Put}$$

in another way, for P = 20% users would on average need a received signal some 7dB louder to overcome the effect of the increased circuit noise. The 20% level was chosen for simplicity of explanation, it must not be assumed that this is a planning figure. Links causing one in five users difficulty of course occur but most are considerably better.

This is all very much simplified but it does remind us that we cannot test everything electronically, communication is a very human affair.

3. NETWORKS

Network analysis and design and the theorems which help in the process are not perhaps the most exciting ingredients in a treatise on communication but some features ought to be examined so that we are not left mystified nor with gaps in our understanding when we study the practical systems which follow.

A network is a combination of one or more generators and electronic components connected together. Of course this can be said of any electric circuit but here we exclude the series-only circuit and limit ourselves to simple basic networks for general application, mainly attenuators and filters. We first meet some theorems which can best be described as ingenious artifices conjured up for us by earlier mathematicians or scientists for making circuit analysis easier.

3.1 THEOREMS

Any simple series circuit is readily solved by Ohm's Law because the current is the same in all parts. When parallel branches are added things get more difficult because currents in the branches create voltage-drops elsewhere in the main circuit but here *Kirchhoff's Laws* come to our aid.

3.1.1 Kirchhoff's Laws

G.R.Kirchhoff (a German physicist) has given us two very helpful reminders by his two laws:

(i) the algebraic sum of the currents meeting at a point in a network is zero. "Algebraic" means that we must be careful about current direction and the law is merely reminding us that the current flowing into a point must be equal to that flowing away from it

(ii) in any closed circuit (called a *mesh*) the algebraic sum of the e.m.f.'s is equal to the algebraic sum of the products of the resistances and their respective currents in the separate parts. In other words, for a mesh, the algebraic sum of the e.m.f.'s and the p.d.'s is zero.

These laws are used to solve a very simple network in the next section but for more complex networks they may lead to several simultaneous equations which absorb much effort in solving, hence the various network theorems which simplify the process by replacing complex networks by lesser ones. As examples, we look at those which are of help to us later in the book.

3.1.2 The Compensation Theorem

This states that any impedance in a network can be replaced by a zero impedance generator having an e.m.f. equal to the instantaneous p.d. which existed across that impedance. In simple terms, an impedance can be replaced by the voltage across it.

We choose a simple circuit to demonstrate the validity and use of the theorem as in Fig. 3.1(i) and in doing so demonstrate the use of Kirchhoff's Laws at the same time. A generator E_1 of, say 1V feeds into the 3-resistor network. Taking the *node* N (point where lines meet), assume a generator current I_1 to be flowing towards it. Now clearly I_1 is the main current which divides between the two resistors R_2 and R_3 so next is shown a current I_2 flowing through R_2, leaving $(I_1 - I_2)$ to flow through R_3. We have in fact been following Kirchhoff's Law No. 1 for at node N, $I_1 - I_2 - (I_1 - I_2) = 0$. Here + is being assigned to currents flowing into the node, – for those flowing away, it does not matter which we use but we must be consistent.

R_2 and R_3 in parallel have a resistance of 10Ω, by Ohm's Law therefore

$$I_1 = \frac{1}{10 + 10} = 0.05\text{A}$$

and because I_1 divides equally between two similar resistances,

$$I_2 = 0.025\text{A} \quad (I_1 - I_2) = 0.025\text{A}$$

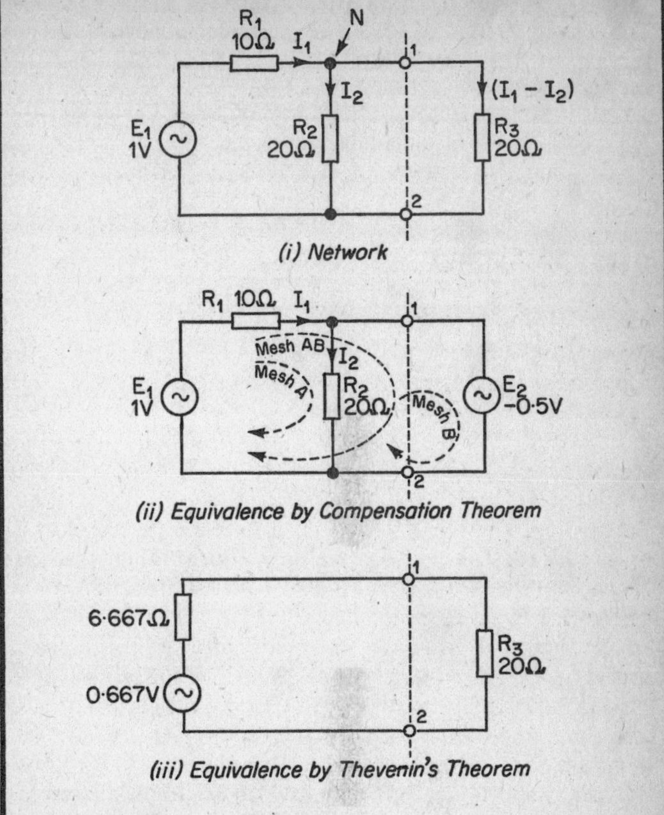

Fig. 3.1 Networks and equivalents

The p.d. across R_3 is 0.025 x 20 = 0.5V and the Compensation Theorem says we can replace R_3 by a zero impedance generator having this p.d., i.e. 0.5V, with a –ve sign because it must be in opposition to E_1 (it could not be +ve unless some voltage is gained from somewhere). This is shown in (ii) of the Figure.

We have next to prove that the theorem works, i.e. that I_1 and

I_2 in the main network are unchanged and this is where Kirchhoff's second law is used.

Mesh AB (E_1, R_1, E_2)

$E_1 + E_2 = I_1 R_1$ (algebraic sum of e.m.f.'s is equal to algebraic sum of p.d.'s)

$$\therefore\ 1 - 0.5 = I_1 \times 10 \quad \therefore\ I_1 = \frac{0.5}{10} = 0.05\text{A as before}$$

Mesh A (E_1, R_1, R_2)

$$E_1 = I_1 R_1 + I_2 R_2 \quad \therefore 1 = (0.05 \times 10) + (I_2 \times 20)$$

$$\therefore\ I_2 = \frac{0.5}{20} = 0.025\text{A as before}$$

or we could have used Mesh B (R_2, E_2) in the direction of the arrow

$$E_2 = I_2 R_2 \qquad \therefore\ -0.5 = -(I_2 \times 20)$$

(I_2 is $-$ve because it flows against the mesh direction we have chosen)

$$\therefore\ I_2 = \frac{0.5}{20} = 0.025\text{A}$$ which agrees with the result using Mesh A.

Thus we have demonstrated the use of Kirchhoff's Laws to solve the network of Fig. 3.1(i) and also shown that by use of the Compensation Theorem an element (R_3 in this case) can be removed and replaced by a generator. The same simple network is used next to demonstrate Thevenin's Theorem.

3.1.3 Thevenin's Theorem

This theorem by M.L.Thevenin (a French engineer) enables us to transform a complicated network into a simple one of a generator plus its internal impedance. In essence the theorem states that if the circuit of Fig. 3.1(i) for example is broken at terminals 1 and 2 then the left hand complex network can be replaced by a generator of voltage and impedance given by the values seen at the terminals looking back into the network. This is most easily demonstrated by a few simple calculations:

The open-circuit voltage looking back into terminals 1 and 2 is given by $I_2 R_2$ (because R_3 is disconnected $I_1 = I_2$).

$$\therefore\ I_2 = \frac{1}{30}\ \text{A} \qquad \therefore\ I_2R_2 = \frac{1}{30} \times 20 = 0.667\text{V}$$

The impedance looking back into terminals 1 and 2 is given by R_2 in parallel with R_1, i.e.

$$\frac{20 \times 10}{20 + 10} = 6.667\Omega$$

hence by Thevenin's Theorem the equivalent circuit can be drawn as in Fig. 3.1(iii) where the network, no matter how complex, has been replaced by an equivalent generator, in this case of 0.667V and internal impedance 6.667Ω. The current in R_3 should not change by this replacement, it is

$$\frac{0.667\text{V}}{(6.667 + 20)\Omega} = 0.025\text{A},$$

the same as found in the above section for the original network.

3.1.4 The Maximum Power Transfer Theorem

Whenever the question of *matching* arises, say between a line and an amplifier or an amplifier and a loudspeaker, it is governed by this theorem. "Matching" means obtaining the electrical condition for maximum power to be transferred from one of the two devices to the other, we will simply refer to them as generator and load.

The theorem states that maximum power is obtained from a generator of internal impedance $Z\angle\phi$ when its load has the complex conjugate impedance (an elaborate way of describing $Z\angle{-\phi}$). Using j notation instead, for a generator of impedance $R + jX$, maximum power is transferred into a load of $R - jX$. However, if the modulus only can be varied, maximum power is obtained when the moduli are equal, irrespective of the value of ϕ.

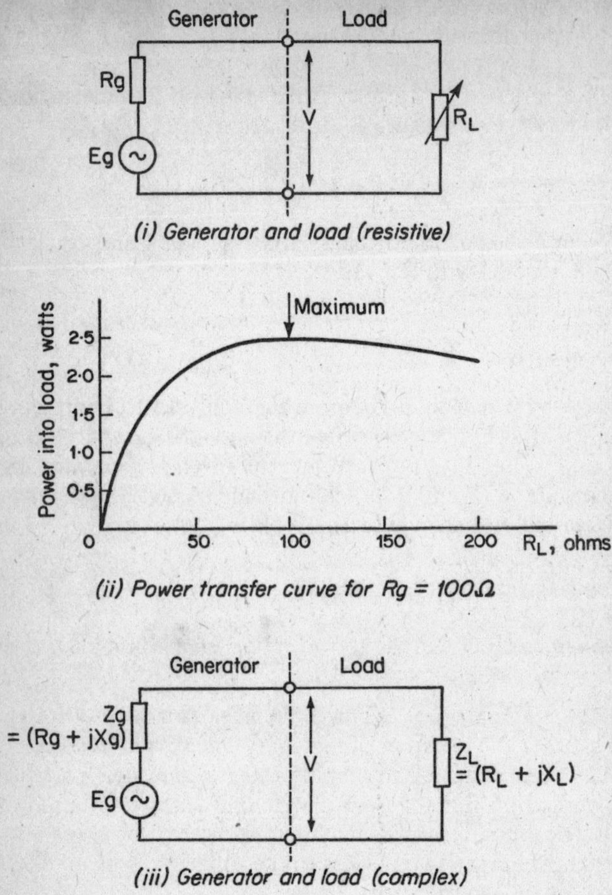

Fig. 3.2 Maximum power transfer

Consider a generator of voltage E_g and internal resistance R_g connected to a variable load R_L as shown in Fig. 3.2(i). We wish to find the value of R_L which produces maximum power transfer from generator to load. Let the generator output voltage = V. Then,

$$V = E_g \times \frac{R_L}{R_L + R_g} \text{ volts}$$

$$\text{and Power in } R_L = \frac{V^2}{R_L} = \frac{E_g{}^2}{R_L} \cdot \frac{R_L{}^2}{(R_L + R_g)^2}$$

$$= E_g{}^2 \cdot \frac{R_L}{(R_L + R_g)^2} \text{ watts.}$$

To see what this means we draw a practical curve for a given value of R_g showing how the power varies as R_L varies, say $R_g = 100\Omega$ with R_L varying from 0 to 200Ω, this is shown in Fig. 3.2(ii). Clearly the power has a maximum value, apparently at $R_L = 100\Omega$, that is when $R_L = R_g$ although it is also evident that considerable mismatch must occur before the power falls significantly from the maximum. Unfortunately we have not quite got the mathematics we need at our disposal for proving this, it would need a whole Appendix, so we generate our own confidence in the validity of the theorem by examining the effect of small variations in R_L around the supposed equality value. Let $R_L = R_g + \delta R_g$ where δ is a fractional change then.

$$\text{Power in } R_L = E_g{}^2 \cdot \frac{R_g + \delta R_g}{(R_g + \delta R_g + R_g)^2}$$

$$= E_g{}^2 \cdot \frac{R_g(1 + \delta)}{R_g{}^2(2 + \delta)^2} = \frac{E_g{}^2}{R_g} \cdot \frac{1 + \delta}{(2 + \delta)^2}$$

$\frac{E_g{}^2}{R_g}$ is constant, we next find what value of δ gives the power maximum value.

δ	0.5	0.1	0.01	0
$\frac{1 + \delta}{(2 + \delta)^2}$	0.24000	0.24943	0.24999	0.25000

δ	– 0.01	– 0.1	– 0.5
$\frac{1 + \delta}{(2 + \delta)^2}$	0.24999	0.24931	0.22222

Undoubtedly the power is maximum when $\delta = 0$ thus the expression $R_L = R_g + \delta R_g$ becomes $R_L = R_g$, for all cases, not just the one illustrated by Fig. 3.2(ii).

Consider next Fig. 3.2(iii) showing generator and load when both are reactive. Through the same reasoning and remembering that no power is dissipated in the load reactance jX_L, it can be shown that:

$$\text{Power in } Z_L = \frac{|V|^2}{R_L} = Eg^2 . \frac{R_L}{(R_L + R_g)^2 + (X_L + X_g)^2}$$

and as far as X_L and X_g are concerned this expression becomes maximum when $(X_L + X_g) = 0$ i.e. $X_L = -X_g$ showing that the reactances must be conjugate so for maximum power transfer in Fig. 3.2(iii)

$$Z_L = R_g - jX_g$$

3.2 NETWORKS

Those we discuss are of the *passive* variety, that is they absorb but do not generate signal power. There are many basic networks, all with different purposes, those which interest us for communication are mainly attenuators which are not frequency-sensitive, to filters which most certainly are. There is in addition one famous network which has impedance balancing as its objective, this is the *Wheatstone Bridge* (after Sir Charles Wheatstone, an English physicist) and we look at this first.

3.2.1 The Wheatstone Bridge

Fig. 3.3(i) shows the basic network, not perhaps in its most recognizable form but so drawn to demonstrate an important feature. There are six impedances interconnected as shown and we shall find that if a generator is connected in place of any one of these, the current in another can be made zero by suitable arrangement of values for the other four, the value of the generator impedance being unimportant. For example, by

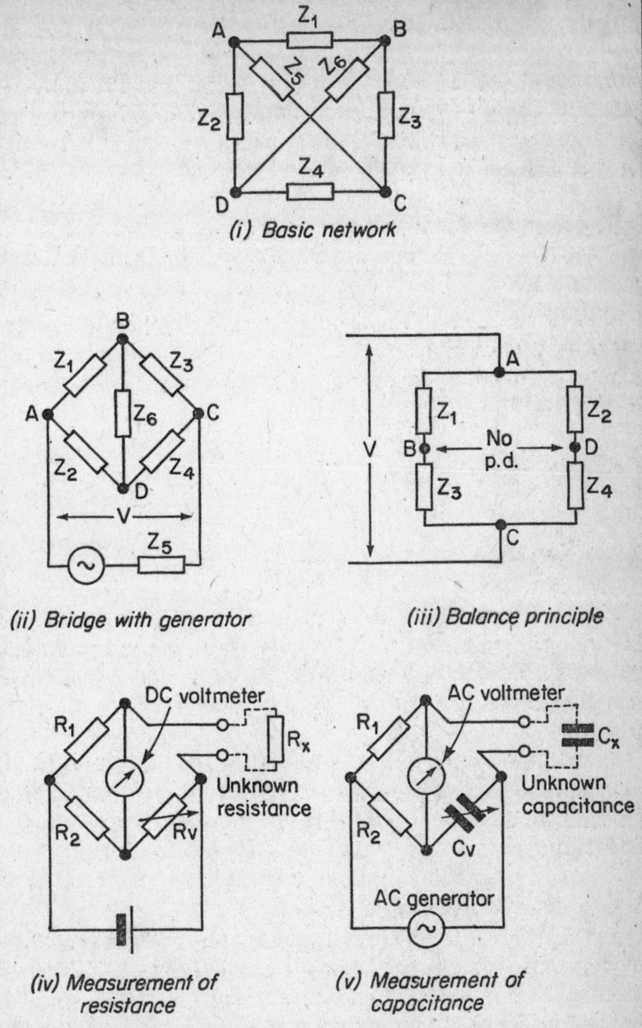

Fig. 3.3 Wheatstone bridge

placing a generator in the Z_5 *arm* and suitably adjusting Z_1, Z_2, Z_3 and Z_4 the current in Z_6 can be made zero and removal of Z_6 will not affect the power absorbed by the network. The circuit with the generator is re-drawn in Fig. 3.3(ii). This is only one example, with the generator in any other arm of the network in (i), a similar circuit to (ii) follows but with the impedances numbered differently, there will always be one in which the current can be made zero. The bridge is then said to be *balanced* and the condition to bring this about can be found through Kirchhoff's Laws or more simply from the voltage-divider principle. For zero current in Z_6 the voltages at nodes B and D must be equal and because the same voltage V is applied to the two impedance chains Z_1 Z_3 and Z_2 Z_4 [Fig. 3.3(iii)]

$$\frac{Z_3}{Z_1 + Z_3} \cdot V = \frac{Z_4}{Z_2 + Z_4} \cdot V$$

$$\therefore\ Z_1Z_4 + Z_3Z_4 = Z_2Z_3 + Z_3Z_4 \quad \therefore Z_1Z_4 = Z_2Z_3$$

or $\dfrac{Z_1}{Z_2} = \dfrac{Z_3}{Z_4}$, the condition for balance.

We have used Z's for impedances but the bridge may comprise resistive or reactive elements only or a combination of both.

Two examples only will suffice to show the usefulness of the Wheatstone Bridge. The simpler is shown in Fig. 3.3(iv) and in this case an a.c. generator is not necessary because the functioning of the bridge can be entirely d.c. For a little realism we use some practical round figures. Let $R_1 = R_2 = 100\Omega$ and R_V be variable over the range $1 - 1000\Omega$ in 1Ω steps.

Then $R_X = R_V \times \dfrac{R_1}{R_2}$ i.e. $R_X = R_V$

so that values of R_X between 1 and 1000Ω can be measured by balancing the bridge (i.e. adjusting R_V for zero deflexion on the meter). Alternatively if both R_1 and R_2 can be switched to 10, 100 or 1000Ω the range of ratios R_1/R_2 is 0.01 to 100, so that values of R_x from 0.01 to 100,000Ω are measurable. This is a practical version of the bridge, the operation of which is also similar if the battery is replaced by an a.c. generator and the d.c. voltmeter by an a.c. type.

Equally the bridge may be arranged for capacitance measurement as in Fig. 3.3(v), the solution of the bridge is

$$C_x = C_v \cdot \frac{R_1}{R_2}$$

C_v is a variable calibrated capacitor and again the $R_1 R_2$ ratios increase the measurement range. There is no frequency term in the formula hence any generator frequency is usable provided that C_x and/or C_v do not vary with frequency.

The whole range of impedances from those with large positive angles (high quality inductors) to those with large negative angles (capacitors) are catered for by various modifications of the basic bridge circuit and even frequency can be measured.

3.2.2 Matching Networks

Having proved the need for matching in Section 3.1.4 it is only right that we should look at methods of its achievement. The simplest form of matching network is the L-section but because this comprises resistances the question immediately arises as to what purpose is served in matching for maximum power transfer if the device used itself creates a loss. We must therefore accept at this stage that matching is also required for other technical purposes and these will become evident as we progress (e.g. Section 6.1.4).

3.2.2.1 The L-Pad

Pad is a term commonly used to describe a network of resistors for attenuation or impedance-matching purposes and the L-pad is the simplest, its configuration is shown in Fig. 3.4(i) where it is considered to be connected between two terminations Z_H and Z_L where Z_H has the higher impedance. We take the mathematics of this particular network step by step to see how it is done, but then avoid the tedium of doing the same for other networks which follow because the basic

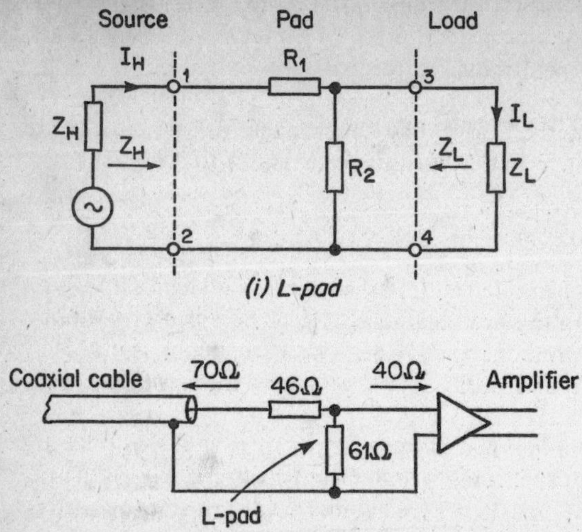

(i) L-pad

(ii) Matching coaxial cable to amplifier

Fig. 3.4 L-type pads

method is similar in each case.

For Z_H to be matched to the terminated network, terminals 1 and 2 must present an impedance Z_H, therefore

$$Z_H = R_1 + \frac{R_2 Z_L}{R_2 + Z_L} \tag{1}$$

Equally at terminals 3 and 4

$$Z_L = \frac{R_2(R_1 + Z_H)}{R_1 + R_2 + Z_H} \tag{2}$$

from (i) $R_2 Z_H + Z_H Z_L = R_1 R_2 + R_1 Z_L + R_2 Z_L$

from (2) $R_1 Z_L + R_2 Z_L + Z_H Z_L = R_1 R_2 + R_2 Z_H$

Adding (1) and (2):

$$R_2 Z_H + 2Z_H Z_L + R_1 Z_L + R_2 Z_L$$

$$= 2R_1 R_2 + R_1 Z_L + R_2 Z_L + R_2 Z_H$$

$$\therefore Z_H Z_L = R_1 R_2 \qquad \text{and} \qquad R_2 = \frac{Z_H Z_L}{R_1}$$

Next, substituting for R_2 in equation (1) [we could equally use (2)]

$$\frac{Z_H{}^2 Z_L}{R_1} + Z_H Z_L = Z_H Z_L + R_1 Z_L + \frac{Z_H Z_L{}^2}{R_1}$$

$$\therefore R_1{}^2 = \frac{Z_H{}^2 Z_L - Z_H Z_L{}^2}{Z_L} = Z_H{}^2 - Z_H Z_L$$

$$= Z_H(Z_H - Z_L)$$

$$\therefore R_1 = \sqrt{Z_H(Z_H - Z_L)}$$

Thus given Z_H and Z_L, the L-pad can be designed.

As an example a need to match a 70Ω coaxial cable to a 40Ω amplifier would require:

$$R_1 = \sqrt{Z_H(Z_H - Z_L)} = \sqrt{70(70 - 40)} \approx 46\Omega$$

$$R_2 = \frac{Z_H Z_L}{R_1} = \frac{70 \times 40}{46} \approx 61\Omega \quad \text{as shown in Fig. 3.4(ii).}$$

These calculations can be quickly checked for the cable should "see" 70Ω and the amplifier 40Ω. For the cable there is 46Ω in series with a parallel combination of 61 and 40Ω, i.e.

$$46 + \frac{61 \times 40}{61 + 40} = 70\Omega$$

and for the amplifier 61Ω in parallel with 70 + 46Ω i.e.

$$\frac{61 \times 116}{61 + 116} = 40\Omega$$

thus giving confidence that the calculations are correct.

Undoubtedly the pad creates a transmission loss because the signal current dissipates power in the resistance arms. In Fig. 3.4(i) let I_H and I_L be the currents in Z_H and Z_L, then

$$I_L = I_H \cdot \frac{R_2}{R_2 + Z_L}$$ (division of current in a 2-resistance parallel network(1/3.4.5))

$$\therefore \frac{I_H}{I_L} = \frac{R_2 + Z_L}{R_2}$$

We must not fall into the trap of thinking that the attenuation in decibels is simply 20 $\log_{10}$ of the current ratio I_H/I_L because these currents flow in different impedances hence

$$\text{attenuation in dB} = 10 \log_{10} \frac{I_H{}^2 Z_H}{I_L{}^2 Z_L}$$

$$= 10 \log_{10} \left(\frac{R_2 + Z_L}{R_2}\right)^2 \cdot \frac{Z_H}{Z_L}$$

and for the example of matching a 70Ω cable to a 40Ω amplifier

$$\text{L-pad attenuation} = 10 \log_{10} \left(\frac{61 + 40}{61}\right)^2 \cdot \frac{70}{40}$$

$$= 10 \log_{10} 4.798 = 6.8\text{dB}.$$

The L-pad in providing a particular match has therefore a certain attenuation which cannot be avoided.

3.2.2.2 *Transformer Matching*

For matching with low attenuation, yet using a passive device, the transformer is inevitable, notwithstanding its high cost compared with that of the few resistors of a pad. *Matching transformers* (as distinct from power transformers) are designed for the particular purpose of matching a source to its load.(2/3.8.3). The important relationship of such a transformer is that the ratio of the two impedances it matches is equal to the square of its *turns ratio*, see Fig. 3.5, i.e

$$\frac{Z_S}{Z_L} = \left(\frac{n_1}{n_2}\right)^2$$

where n_1 and n_2 are the numbers of turns connected to Z_S and Z_L respectively, hence n_1/n_2 is the turns ratio.

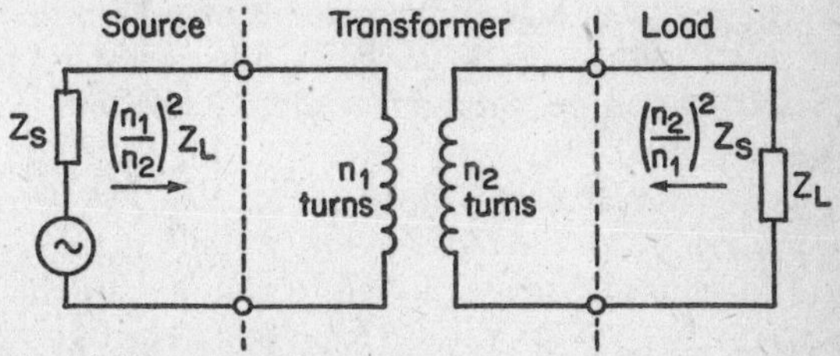

Fig. 3.5 Matching transformer

For example, if it is required to match a 1200Ω line (Z_S) to a 600Ω line amplifier (Z_L):

$$\frac{1200}{600} = \left(\frac{n_1}{n_2}\right)^2 \qquad \therefore \frac{n_1}{n_2} = \sqrt{2} = 1.414$$

i.e. the turns ratio of the two windings is 1.414 : 1. This only determines the ratio, the actual numbers of turns used in a practical design depend on many other factors such as frequency range, core, etc., usually a compromise to keep losses at a minimum. Generally a line matching transformer of this type has a loss of some 0.5 – 1.0dB.

3.2.3 Attenuating Networks

Although attenuation is usually a hindrance to communication, there are occasions when we deliberately introduce it, the loudness control on a radio receiver is an example. A less well-known one but more in keeping with this section is the technique frequently used with line amplifiers in which a standard negative feedback amplifier of fixed gain (say, 30dB) is used and lower gains obtained by wiring an attenuator of the required loss into the input or output terminals. Such an attenuator works between equal impedances, the pad itself is therefore *symmetrical* as we shall see in the next section. Containing resistances only, its loss is mainly independent of frequency.

3.2.3.1 The Symmetrical T-Pad

Because such a pad when inserted between two equal impedances must not disturb the matching between them it is evident that for a pad to be designed, not only must the desired attenuation be quoted but also the termination impedance. A pad of this type is shown in Fig. 3.6(i), here it is connected between a source Z_o and a load Z_o. The two series arms are equal, hence the term "symmetrical". If the source and load currents are labelled I_S and I_L and the ratio between them, $I_S/I_L = N$, then for a pad attenuation of αdB,

since $\alpha = 20 \log_{10} \frac{I_S}{I_L}$, $\alpha = 20 \log_{10} N$ and $N = \text{antilog} \frac{\alpha}{20}$.

The design formulae are now simplified because by working with N rather than α, logarithmic units in the expressions are avoided, they are:

$$R_1 = Z_o\left(\frac{N-1}{N+1}\right) \qquad R_2 = Z_o\left(\frac{2N}{N^2-1}\right)$$

so that for example, for a 10dB attenuator to work between 600 ohm ($\angle 0°$) impedances:

$$N = \text{antilog}\ \frac{10}{20} = 3.162$$

$$R_1 = 600 \cdot \frac{2.162}{4.162} \approx 312\Omega$$

$$R_2 = 600 \cdot \frac{6.324}{8.998} \approx 422\Omega$$

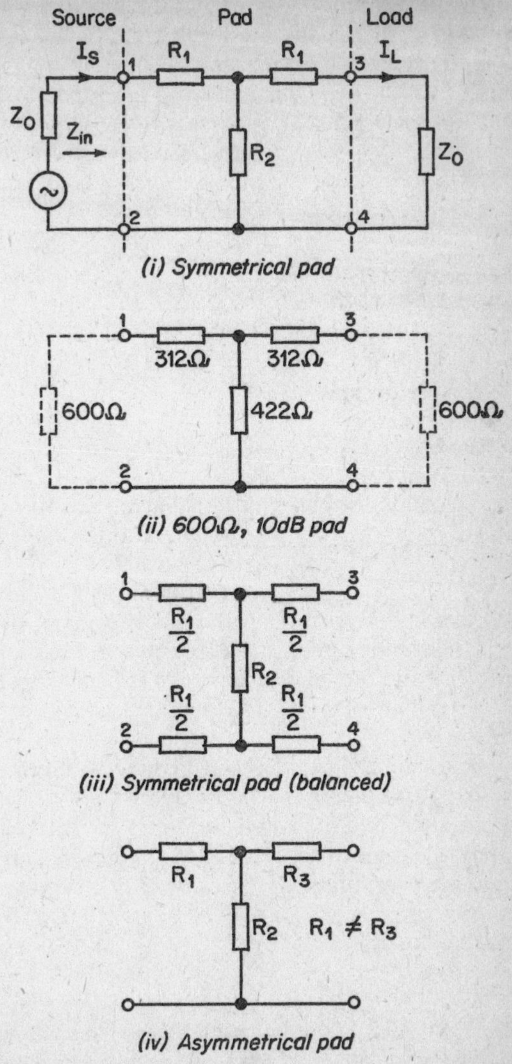

Fig. 3.6 T-type pads

The attenuator is shown in Fig. 3.6(ii) and it only creates 10dB loss when connected between 600 ohm impedances. Terminals 1, 2 and 3,4 may be changed over because the pad is symmetrical.

The impedance of the pad when correctly terminated, i.e. looking into terminals 1 and 2 of Fig. 3.6(i), is

$$Z_{in} = R_1 + \frac{R_2 (R_1 + Z_o)}{R_1 + R_2 + Z_o} \qquad \text{but } Z_{in} = Zo$$

$$\therefore Z_o = \frac{R_1^2 + 2R_1 R_2 + R_1 Z_o + R_2 Z_o}{R_1 + R_2 + Z_o}$$

from which

$$Z_o = \sqrt{R_1^2 + 2 R_1 R_2}$$

and for the pad in the above example [Fig. 3.6(ii)]

$$Z_o = \sqrt{312^2 + 2 \times 312 \times 422} \approx 600\Omega$$

showing that insertion of the pad, while creating a 10dB loss does not change the impedance connected to the source, nor that connected to the load because of the symmetry of the whole circuit.

The T-pad so far considered is said to be *unbalanced* because the series resistors are in one wire only of the through circuit, usually used when the 2 and 4 terminals are earthed or connected to chassis or common.(3/2.2) Where this is not so and a *balanced* version is required, Fig. 3.6(iii) applies in which each series resistance (R_1) is divided equally between the two wires of the circuit.

3.2.3.2 The Asymmetrical T-Pad

From Section 3.2.2.1 it is evident that the L-pad in matching two unequal impedances incurs a certain value of loss. If therefore an L-pad is converted into an asymmetrical T-pad [Fig. 3.6(iv)] by the addition of one extra series resistor (R_3) we then have a matching pad which at the same time introduces some specified value of loss. However at a certain minimum value of loss we are back to the L-pad again because R_3 has been reduced to zero. The asymmetrical T-pad is equally convertible to the balanced form.

3.2.3.3 The π-Pad

Occasionally a different pad configuration is useful in that the source and load first meet a shunt arm instead of a series (often useful when d.c. is carried). This is the π-pad which has one series resistor flanked by two shunt ones. The techniques of analysis we have used with T-pads apply equally to the π-type but of course resulting in modified formulae for calculating component values. The π-pad is obtainable symmetrical or asymmetrical, balanced or unbalanced and a well-known set of formulae relate T and π-pads so that we can, for example, design a T-pad and convert it to the equivalent π to work between the same impedances while introducing the same loss.

3.3 FILTERS

The theoretical considerations in the design of filters, especially those used in multiplex telephony systems which we meet in the next chapter, are complicated. Hence, although our level of mathematical ability should now be such that we could gain some fair insight into the functioning and design of filters, this is not the place to do so as our aim tends towards the elements of systems and techniques rather than of the components. Accordingly we will mainly confine ourselves in this section to the characteristics of the three elementary types.

As far as frequency is concerned filters are the opposite of attenuators for whereas the latter should have no change in attenuation with frequency, the filter has and usually quite sharply. We label the frequency at which the attenuation changes, f_c and the three basic filter types are:

(i) *low-pass* for which the attenuation α is theoretically zero up to f_c whereupon it rises rapidly as frequency changes above f_c so giving a *pass band* from $0 - f_c$ Hz and an *attenuation band from* $f_c - \infty$ Hz. The ideal characteristic is given in Fig 3.7 (i).

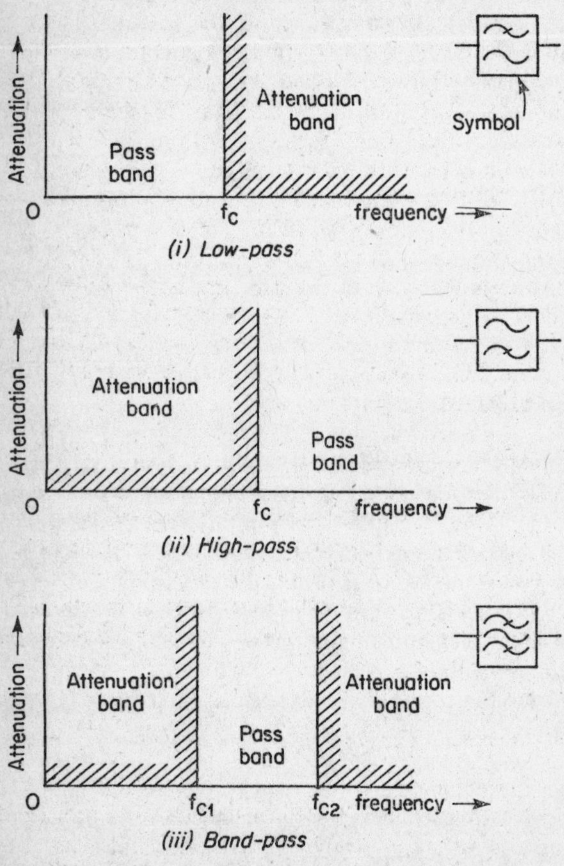

Fig. 3.7 Ideal filter characteristics

(ii) *high-pass* – the attenuation is high up to f_c then falls rapidly to zero so giving an attenuation band $0 - f_c$ Hz and pass band $f_c - \infty$Hz [Fig. 3.7(ii)]

(iii) *band-pass* – ideally as shown in Fig. 3.7(iii), there is a pass band between f_{c_1} and f_{c_2} with attenuation bands above and below.

The general circuit symbols are shown in the figure.

Not unreasonably one might expect inductance and capacitance to play a major part in frequency selection and this is so and with the equivalent in crystal filters, however I.C. operational amplifiers(3/4.3.1.1) can dispense with the bulky inductor and provide all the electrical characteristics of LC networks by using resistors and capacitors only. Much filter design needs computer help.

As with attenuators filter sections are designed to work between known terminating impedances which are related to a design parameter R_o, known as the *design impedance*. The sections may be in either T or π form and may be used in tandem to increase the sharpness of cut-off.

3.3.1 Low-Pass Sections

Simple T and π low-pass filter sections are shown in Fig. 3.8(i), there is a total series inductance of L and shunt capacitance C in both cases. As the frequency rises the series reactance increases and the shunt reactance falls (2/3.1) the loss therefore increases with frequency although not proportionally as with a single reactance, there is a relatively sudden change. For the low pass filter.

$$f_c = \frac{1}{\pi\sqrt{LC}} \text{ Hz}, \quad L = \frac{R_o}{\pi f_c} \text{ H}, \quad C = \frac{1}{\pi f_c R_o} \text{ F}$$

and as an example, for a 4kHz low-pass filter with $R_o = 600\Omega$

$L = 47.75$mH, $\quad L/2 = 23.87$mH, $\quad C = 0.133\mu$F,

$$\frac{C}{2} = 0.0663\mu\text{F}$$

and the two types of filter section are shown in Fig. 3.8(ii) with the approximate α/f characteristic in (iii). Because this is a single simple filter section the rise in attention at f_c falls very short of the ideal, for example at 5kHz (1.25 x f_c), α is only 12dB. Very much higher values are normally desirable and are obtainable with more sections (e.g. for two sections α at 5kHz = 24dB) or by use of more complex filter networks.

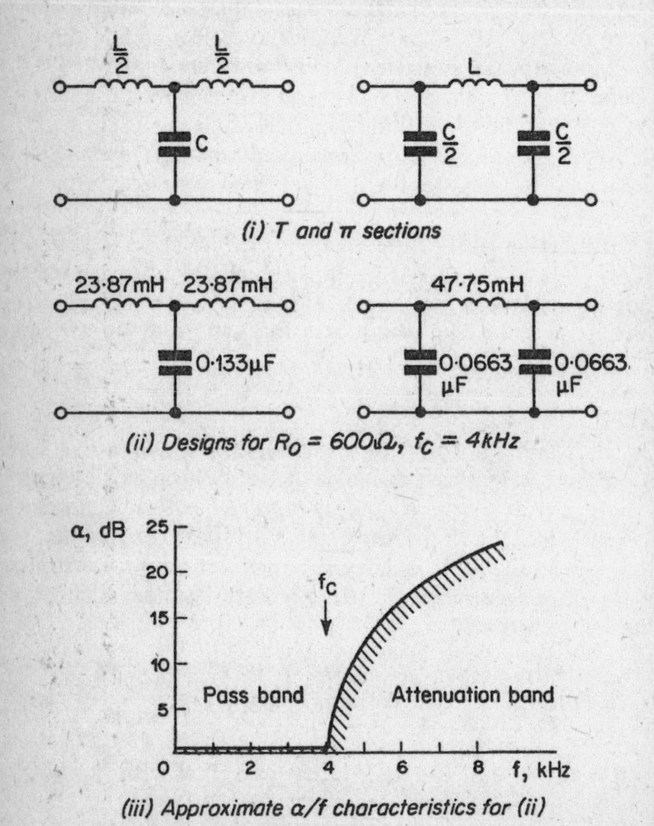

Fig. 3.8 Low-pass filter sections

Some small attenuation is inevitable in the pass-band because of resistance losses, especially in the inductor, these also add up when more than one section is used.

3.3.2 High-Pass Sections

The filter and its design run complementary to that of the low-pass section, the components are arranged as in Fig. 3.9(i).

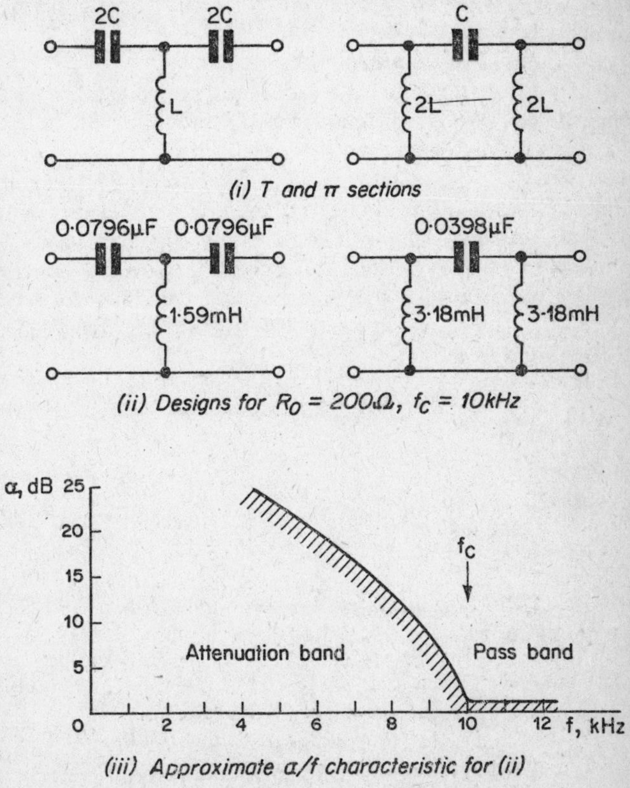

Fig. 3.9 High-pass filter sections

As frequency rises the series reactance falls while the shunt reactance rises, the conditions for lower attenuation at higher frequencies. Design formulae are as follows:

$$f_c = \frac{1}{4\pi\sqrt{LC}} \text{ Hz}, \quad L = \frac{R_o}{4\pi f_c} \text{ H}, \quad C = \frac{1}{4\pi f_c R_o} \text{ F}$$

and as an example, for a 10kHz high-pass filter with $R_o = 200\Omega$, $L = 1.59$mH, $2L = 3.18$mH, $C = 0.0398\mu$F, $2C = 0.0796\mu$F and the two types of filter section are shown in Fig. 3.9(ii) with the approximate α/f characteristic in (iii).

3.3.3 Band-Pass Sections

Evidently a requirement for a band-pass filter having a pass band between two given frequencies f_{c_1} and f_{c_2} could be met by design of a high-pass filter cutting off at f_{c_1} followed or preceded by a low-pass cutting off at f_{c_2}. Alternatively a complete band-pass section can be produced by a similar procedure as for the separate filters, it is naturally slightly more complicated because of the duplication of cut-off frequencies. Basically the section contains series and parallel resonant circuits(2/3.7) as shown in Fig. 3.10(i). The component values are calculated from:

$$L_1 = \frac{R_o}{\pi(f_{c_2} - f_{c_1})} \qquad C_1 = \frac{(f_{c_2} - f_{c_1})}{4\pi f_{c_1} \cdot f_{c_2} \cdot R_o},$$

$$L_2 = \frac{(f_{c_2} - f_{c_1}) R_o}{4\pi f_{c_1} \cdot f_{c_2}} \qquad C_2 = \frac{1}{\pi(f_{c_2} - f_{c_1}) R_o}$$

Suppose a filter is required with a pass-band from 4 – 8 kHz and design impedance 600Ω, then

$f_{c_1} = 4$kHz, $f_{c_2} = 8$kHz, $R_o = 600\Omega$ and

$L_1 = 47.75$mH, $L_1/2 = 23.87$mH, $L_2 = 5.97$mH

$2L_2 = 11.94$mH, $C_1 = 0.0166\mu$F $2C_1 = 0.0332\mu$F,

$C_2 = 0.133\mu$F, $C_2/2 = 0.0663\mu$F,

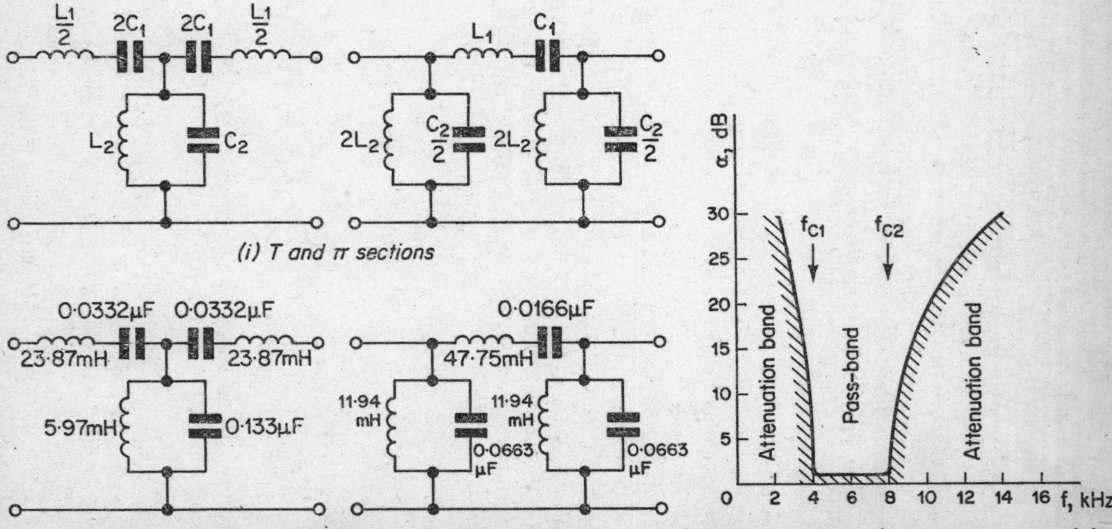

(i) T and π sections

(ii) Designs for $R_0 = 600\Omega$, $f_{C1} = 4$kHz, $f_{C2} = 8$kHz

(iii) Approximate α/f characteristic for (ii)

Fig. 3.10 Band-pass filter sections.

resulting in the practical design as shown in Fig. 3.10(ii) which has an α/f characteristic as in (iii).

An interesting feature of this type of design which we will appreciate better when we have studied multi-channel systems, is that because each channel requires a filter of the same bandwidth, L_1 and C_2 are the same for all filters. The formulae show that they do not depend on the position of the channel in the system frequency band, only on the actual channel bandwidth which is the same for all channels in the system.

4. TRANSMISSION SYSTEM TECHNIQUES

Chapter 1 introduces us in a general way to the idea of a certain bandwidth being required for a signal to carry successfully the intended information. We will study the actual bandwidths for various types of signal as we progress but what we must first begin to appreciate are the various techniques by which the bands of frequencies which make up signals are "carried".

Of the two main methods of signal transmission, line and radio, the line offers the simpler because it can carry many types of signal *directly*. For example, microphone signals may be transmitted directly over a two-wire line (usually a pair of wires in an underground cable or perhaps overhead on poles) which for moderate distances has adequate bandwidth. The attenuation unfortunately is not constant with frequency because of cable capacitance. The effect is not so pronounced with overhead wires because of their greater spacing and air dielectric.[1/4.2.2] However:

(i) as distance increases the attenuation causes the signal level to fall until ultimately the weakness of the signal and the poor signal/noise ratio render communication impossible
(ii) generally we just cannot have an exclusive pair of wires for each channel, especially for long-distance circuits.

The techniques which follow overcome these restrictions but not without a few problems.

4.1 AUDIO FREQUENCY SYSTEMS

Retrieval of an attenuated signal is usually accomplished through amplification but there is one alternative which may be classed as such but which actually raises the signal level by presenting a *negative impedance*, a concept which may be found difficult to accept. Mathematically it is sound because if there is a loss when a signal flows through a positive impedance, it is reasonable to expect a gain with a negative impedance. The difficulty lies in visualizing a negative impedance. It is in fact obtained by using transistor circuits with positive feedback and it can well be imagined that stability presents problems.(3/3.2.7) For this and other complex reasons the gain can be no greater than about 10 – 12dB, so limiting the use to relatively short distances, say 20 – 30km. Nevertheless the fact that a negative impedance amplifier can be inserted in a 2-wire line and does not need 4 wires (or 2 channels) as do most other systems means that it has the great advantage of lower line cost. It can be considered as equivalent to a T-network (Section 3.2.3) comprising both series and shunt negative impedance elements.

We next examine the more conventional 2-wire amplified circuit because from this we learn so much about the fundamentals of amplified line systems as a whole.

4.1.1 The Two-Wire Amplifier

Unlike the negative impedance amplifier which being simply in the form of a T-network is a both-way device, the single amplifier(3/3.2) is one-way. It has input and output terminals, a signal applied to the input appears amplified at the output but amplification does not take place in the reverse direction. Fig. 4.1(i) shows the negative-impedance and an elementary conventional amplifier method together, the latter is technically practical and the amplifier gains might reasonably be adjusted so that the speech sound pressure delivered by each receiver is approximately the same as that reaching the microphone at the sending end of the channel. However, this is a 4-wire system throughout and to reduce it to 2-wire, a means of combining the transmit and receive paths on both sides of the amplifiers is required.

4.1.1.1 Terminating Sets

A direct connexion between the two pairs of wires is unusable because it also results in the connexion of the output of each amplifier to the input of the other so forming a complete positive feedback system which is unstable,(3/3.3) see

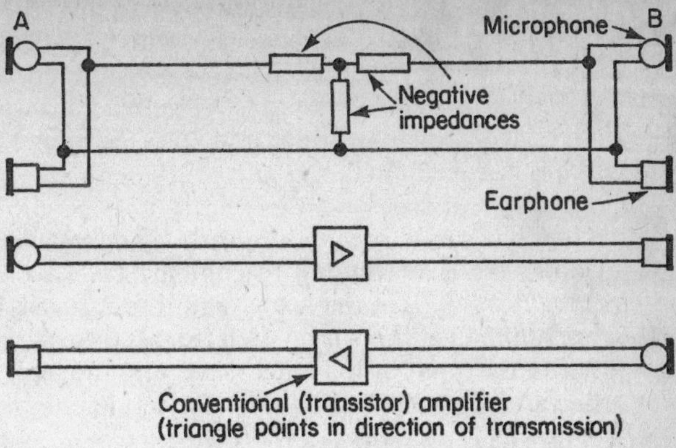

(i) Comparison of negative impedance and conventional amplifier systems

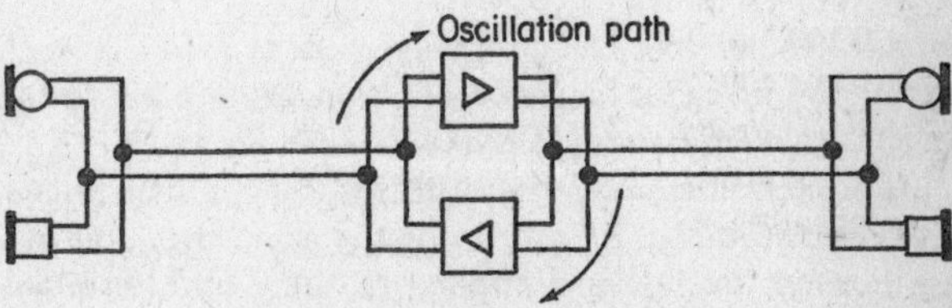

(ii) Unworkable 2-wire system.

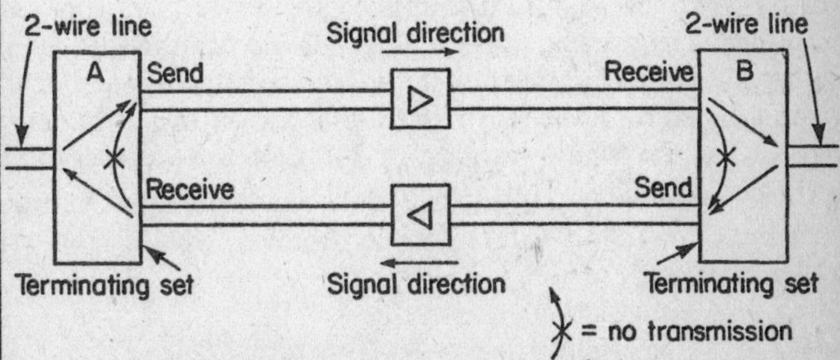

(iii) Transmission paths in 4w/2w terminating sets

Fig. 4.1 Development of terminating set

Fig. 4.1(ii). Combining the two paths into a single one without oscillation is accomplished by use of a special resistive terminating network or by a *hybrid transformer*. Although the latter is more likely to be used, the resistive network brings out the underlying Wheatstone Bridge principles (Section 3.2.1) and therefore is discussed first.

The design problem can be seen more clearly by reference to Fig. 4.1(iii). We are endeavouring to combine the two amplifiers in such a way that the A to B and B to A signals are amplified as required yet there is no path from the output of one amplifier to the input of the other. Note that the send path at A becomes the receive path at B. The terminating sets shown therefore have six terminals for 2-wire line, send and receive paths. Simply stated, the set must provide signal paths 2-wire to Send, Receive to 2-wire, but block transmission from Receive to Send.

A suitable Wheatstone Bridge arrangement is as in Fig. 4.2(i). Assuming the bridge is balanced, if a signal is applied to the Receive terminals, currents flow in Z_1, Z_2, Z_4 and the 2-wire line but none in the Send circuit, this is the condition needed for amplifier stability. By redrawing the same circuit as in (ii) it can be seen that a signal applied to the 2-wire terminals produces current in Z_1, Z_4, Send and Receive. The current in Send is essential, that in Receive serves no purpose but equally has no ill effect because it is applied to the *output* of an amplifier. The arrangements of Fig. 4.2 are therefore satisfactory but only provided that the bridge is balanced. It is when a signal is on the receive path that conditions are most critical for the whole purpose of the unit is to prevent amplifier instability. Thus from Fig. 4.2(i), for balance

$$\frac{Z_2}{Z_4} = \frac{Z_1}{Z_{(2w)}}$$

If Z_1 and Z_2 are for example, made equal then Z_4 must be equal to the impedance of the 2-wire line, achieving this is the limiting factor.

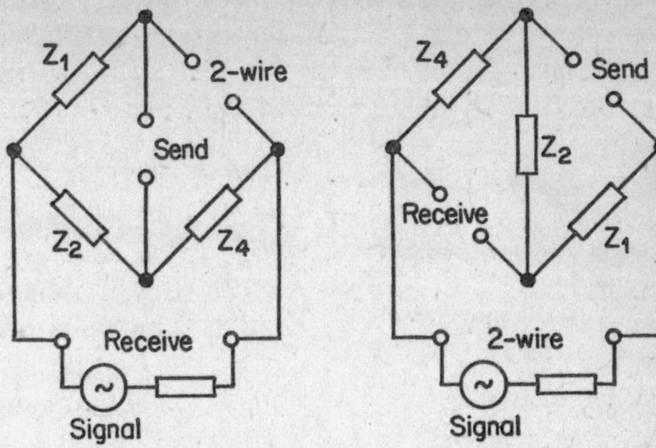

(i) Signal on receive terminals

(ii) Signal on 2-wire terminals

Fig. 4.2 Wheatstone bridge arrangements for 4w/2w terminating set

Now, resistances absorb power, therefore the more efficient terminating set employs transformer windings instead. Consider the basic hybrid transformer circuit of Fig. 4.3(i), ("hybrid" because it is a mixture of three windings). The difference from Fig. 4.2(i) is simply that Z_2 and Z_4 have been replaced by transformer windings coupled to a third winding connected to the receive circuit which is now represented by its impedance Z_R. Assuming that windings 1 and 2 of the hybrid transformer are connected so that the network currents i_1 and i_2 flow in the directions shown, then if they are equal and in opposition in the central branch Z_S (the impedance of the sending path), they cancel. For i_1 and i_2 to be equal and assuming that windings 1 and 2 have equal e.m.f.'s induced in them, Z_B must be equal to Z_L, the impedance of the 2-wire line. Z_B is called the *balance impedance*. Thus we have similar conditions and requirements as in Fig. 4.2 without the losses of Z_2 and Z_4, a preferable but more expensive arrangement.

This is how the circuit should be remembered, by referring to

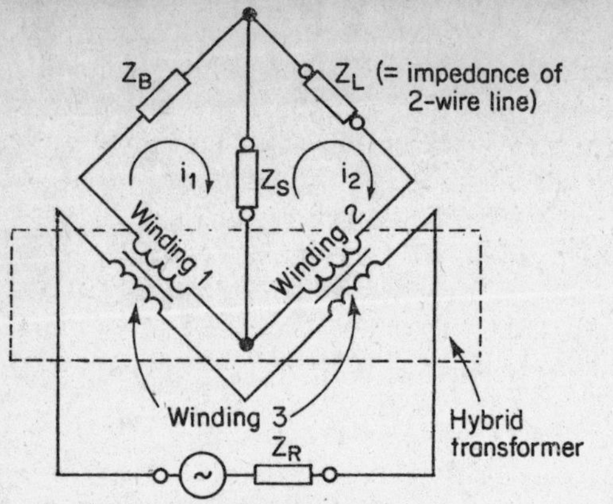

(i) Drawn as a Wheatstone bridge

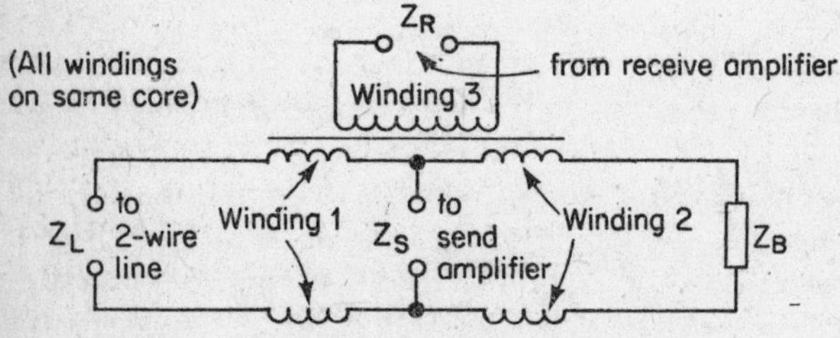

(ii) Balanced circuit

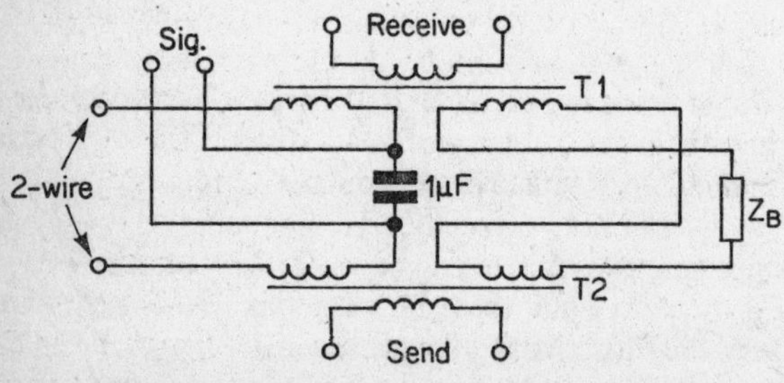

(iii) Telephony circuit using two transformers

Fig. 4.3 4w/2w terminating set

the Wheatstone Bridge for by so doing the principles are more obvious. In practice, because circuits need to be balanced electrically [as for the T-type pad in Fig. 3.6(iii)], windings 1 and 2 are each split and connected into both wires of the appropriate branch of the circuit as shown in Fig. 4.3(ii). A complete telephony circuit using two separate transformers is shown in Fig. 4.3(iii). It includes a metallic path through the unit from 2-w to SIG. This carries d.c. signals for switching or signalling. The capacitor (1 or 2μF) prevents them being short-circuited but presents a relatively low reactance to the a.c. signal. We will see how these terminals are connected later.

If the microphones and receivers of Fig. 4.1 are those of a telephone instrument, that is, together in a *handset*, they too must be coupled through a hybrid transformer to the 2-wire line. The circuit principles are the same as developed above, in this case the microphone must deliver speech power to the line but not to the receiver (if we hear ourselves too loudly we lower our voices) and in the other direction the line must deliver power to the receiver. We can now demonstrate the use of terminating sets by drawing a complete amplified 2-wire line circuit as in Fig. 4.4. Such circuits are in common use with lines of attenuation of up to some 10dB (say, about 20 – 40km, depending on the wire gauge used). The SIG terminals are connected straight across the amplifier system to give a continuous d.c. path from end to end. Signalling equipment is not shown but in its most elementary form needs only to comprise a press-button, bell and battery at each end.

Before looking separately at the principles of line impedance balancing, we sum up the main transmission characteristics of the terminating set (it may help to refer to Fig. 4.2).

(i) When a signal appears on the 2-wire line terminals, approximately the same signal power is wasted in the output of the receive amplifier as is usefully used in the send amplifier, the loss from 2-wire to Send must there-be at least 3dB, allowing for losses in the transformer, say 3.5 – 4.0dB. This loss is easily regained by the amplifier.

(ii) the path from Receive to 2-wire has a similar loss as in

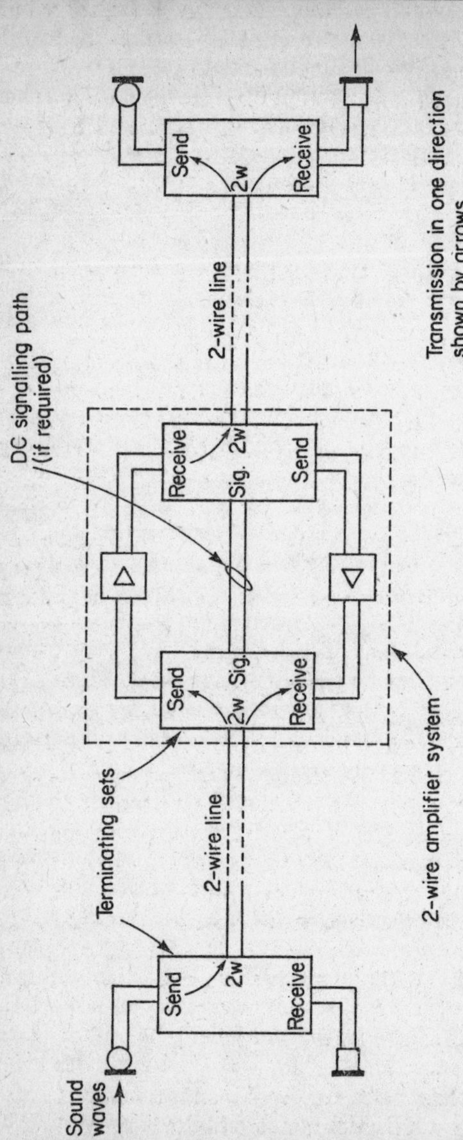

Fig. 4.4 2-wire amplified circuit

(i) because as much power is dissipated in the balance as in the line. Again the amplifier restores the lost power.

(iii) the degree to which unwanted power is fed from Receive to Send is controlled by the electrical balance between the impedance of the line and that of the balancing network. Because instability renders the system useless, it is this balance which sets the limit to the total amplifier gain usable.

Line amplifiers are frequently known as *repeaters*.

4.1.1.2 Balance Networks

Notwithstanding the complication of two amplifiers and two 4w/2w terminating sets in a 2-wire amplifier system, a gain of only about 10dB is realized. Higher overall gains are possible provided that Z_B and Z_L of Fig. 4.3(ii) are well matched. We must not forget that we are dealing with a range of frequencies, for speech for example, some 3kHz wide and the two entities need to match over the whole range. Consider the amplifier system of Fig. 4.4 as redrawn in Fig. 4.5 to show gains and losses round the *loop*. The two sides of the loop are labelled *Near-End* (n.e.) and *Far-End* (f.e.), we are at the near-end hence the top amplifier is in the send direction while the bottom one is for receiving. Unless the balancing is perfect, power is transmitted from the Receive terminals of each terminating set to the Send terminals. The attenuation between the two pairs of terminals is called the *transhybrid loss*. This has two components, a relatively constant one of 6dB (made up of the power delivered to the line and that lost in the remainder of the hybrid circuit) and a variable one known as the *return loss* (RL). More precisely the transhybrid loss is defined as the ratio in decibels of the received power to that transmitted and is equal to (6 + RL)dB. By clever use of Chapters 2 and 3 it is possible to obtain the expression

$$\text{Return Loss} = 20\log_{10}\left|\frac{Z_B + Z_L}{Z_B - Z_L}\right| \text{ dB}$$

Before looking at this in detail, a reminder of the problem. In Fig. 4.5 the total amplifier gain in the loop is $(A_s + A_r)$ and the total attenuation $(\alpha_{ne} + \alpha_{fe})$, all quoted in decibels. For stability ($A_s + A_r$) must not exceed ($\alpha_{ne} + \alpha_{fe}$).

If the lines and transducers are unlikely to change, a reasonably precise balance network can be designed. Lines usually are capacitive so a resistance-capacitance network is required. However individual balancing is expensive in design and installation effort so to make terminating sets available for more general use a *compromise* balance is usually built in, "compromise" meaning that it does its best considering the large range of 2-wire line impedances the set may have to meet. Compromise balances are frequently resistors of 600 or 900Ω.

To get some first-hand experience with this, we put into practice the a.c. theory of Book 2 and calculate the total allowable gain for a 2-wire amplifier as in Fig. 4.5. This not only gives practice but because terminating sets appear in so many types of communication circuit, it is important that we get to grips with them fully right from the beginning.

Calculations involving impedances contain many pitfalls

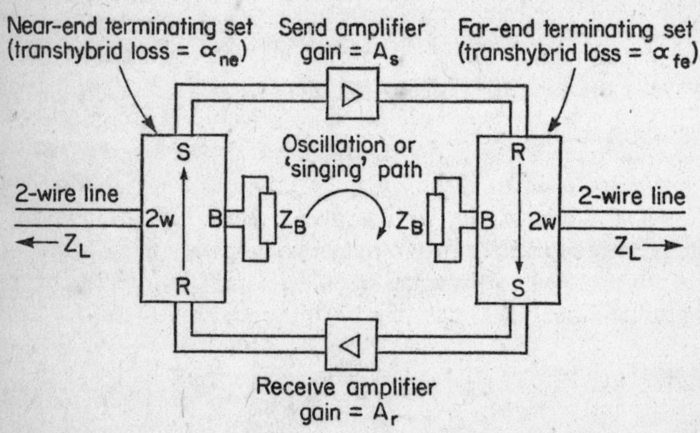

Fig. 4.5 Two-wire amplifier system

besides usually involving large numbers so to make things relatively simple yet to be able to feel that we have had some involvement with the process, we consider a frequency range of 500 – 3000 Hz only and that the near-end and far-end 2-wire lines have the same impedance, a not unreasonable assumption when the amplifier system is installed at the centre of the line. By disconnecting the 2-wire line terminals from the terminating set and using an impedance measuring set across the line we obtain the following readings:

At 500 Hz	$Z_L = 900 - j450$
At 1000 Hz	$Z_L = 600 - j500$
At 2000 Hz	$Z_L = 360 - j430$
At 3000 Hz	$Z_L = 250 - j350$

(these are practical values, modified to demonstrate the principles more vividly – note the negative j term in each case because the line is predominantly capacitive).

Consider a compromise balance in each terminating set of 600Ω, then at 500 Hz

$$Z_B = 600 + j0,$$

$$Z_L = 900 - j450 \quad \text{and}$$

$$\text{Return loss} = 20\log_{10}\left|\frac{Z_B + Z_L}{Z_B - Z_L}\right| \text{ dB},$$

and we recall that the vertical lines mean that we must calculate the *modulus* of the net impedance

$$\therefore \text{ Return loss} = 20 \log \left|\frac{600 + 900 - j450}{600 - 900 + j450}\right|$$

$$= 20 \log \left|\frac{150 - j45}{-30 + j45}\right|$$

$$= 20 \log \left|\frac{(150 - j45)(-30 - j45)}{(-30 + j45)(-30 - j45)}\right|$$

$$= 20 \log \left| \frac{-4500 - j6750 + j1350 - 2025}{(-30)^2 - (j45)^2} \right|$$

$$= 20 \log \left| \frac{-6525 - j5400}{900 + 2025} \right|$$

$$= 20 \log \left| -2.23 - j\,1.85 \right|$$

The modulus of (a + jb) is given by $\sqrt{a^2 + b^2}$, therefore

$$\text{Return loss} = 20 \log \sqrt{2.23^2 + 1.85^2} = 20 \log 2.897$$

$$= 20 \times 0.4619$$

$\therefore$ Return loss = 9.2dB.

By repeating the calculations at the other frequencies we get:
Return loss with $Z_B = 600\Omega$

Hz	500	1000	2000	3000
dB	9.2	8.3	6.6	5.4

This is for each terminating set, therefore the total attenuation ($\alpha_{ne} + \alpha_{fe}$) at each frequency is double the above figure plus 12dB. The minimum attenuation is at 3000 Hz, 2 (6 + 5.4) = 22.8dB and ($A_s + A_r$) must not exceed this value otherwise the loop will oscillate at or near this frequency. Thus with ($A_s + A_r$) $\approx$ 22.8dB, this could usually be apportioned as about 11dB gain for each amplifier. (note – in this brief look at the techniques, we ignore frequencies above 3000 Hz but in practice should the attenuation be even less at any of these, from Section 3.3.1 it is evident that a LP filter with cut-off at 3000 Hz somewhere in the loop would raise the attenuation above but not below 3000 Hz, such a filter is shown in use in Fig. 4.7).

The attenuations calculated are for a compromise balance. A much better match between Z_L and Z_B is given when the latter is a simple network of a resistance of 1050Ω in parallel with a capacitance of 0.13μF. We again calculate the allowable ($A_s + A_r$) using such a network. Firstly its impedances are required at the frequencies being considered and we recall that for a parallel network:(3/3.6)

$$\frac{1}{Z_T} = \frac{1}{Z_1} + \frac{1}{Z_2}$$

where Z_1 and Z_2 refer to the two branches of the network and Z_T is the total impedance.

Here $Z_1 = R$, $Z_2 = \frac{-j}{\omega C}$ (or $\frac{1}{j\omega C}$ by multiplying by $\frac{j}{j}$)

$$\therefore \frac{1}{Z} = \frac{1}{R} + \frac{1}{\frac{1}{j\omega c}} = \frac{1}{R} + \frac{j\omega c}{1} = \frac{1 + j\omega CR}{R}$$

$$\therefore Z = \frac{R}{1 + j\omega CR}$$

At 500 Hz. R = 1050Ω $\omega C = 2\pi \times 500 \times 0.13 \times 10^{-6}$

$$\therefore Z = \frac{1050}{1 + j0.4288} = \frac{1050\,(1 - j0.4288)}{1^2 + 0.4288^2}$$

$$= \frac{1050 - j450}{1.1839} \approx 890 - j380$$

which is the new value for Z_B. Then by calculating the return loss we obtain a value of 28.9dB, a considerable improvement over 9.2dB for the compromise balance at this frequency. Repeating the calculations for the remaining frequencies gives the whole picture:

Return loss with R = 1050Ω in parallel with C = 0.13μF:

Hz	500	1000	2000	3000
dB	28.9	37.7	20.8	17.3

so our new balance enables ($A_s + A_r$) to be as much as

$2(6 + 17.3) \approx 46$dB compared with about 22dB for the 600Ω compromise balance. The special network actually costs very little more. A picture in the form of an impedance diagram as in Fig. 4.6 shows what has happened. The greater the distance on the graph between the balance and line impedances (as shown for 3000 Hz), the lower the return loss. What is most significant in this exercise is that for the mere cost of the small capacitor, the gain available from the system is increased by about 12dB in both directions, equivalent to the transmission loss of several kilometres of underground cable. An exaggerated case perhaps but still a practical one.

In long-distance telephony circuits such precision balancing is unusable because the 2-wire line switched into use on a particular call could be one of many thousands available, all varying in length and wire gauge, hence the balance can only be a compromise.

4.1.2 Four-Wire Circuits

When more gain is required than is available from a 2-wire amplifier as developed in the previous section, 4-wire working is employed. Having studied the 2-wire system in some depth, we shall have little difficulty with the 4-wire because it is simply an extension of the loop of Fig. 4.5 to contain more lines and amplifiers as shown in Fig. 4.7. The balancing principles of the terminating sets and the restriction that total gain must not exceed total loss within the loop are unchanged. Spacing between amplifiers can be up to 50 – 60 km, the main requirement being that amplification must be provided before the signal/noise ratio has deteriorated excessively. For commercial speech this ratio should be at least 40dB. There is no limit to the length of such circuits except that *echo* may become troublesome. Echo is the return of a signal from an imperfectly balanced terminating set at the far-end (most are), delayed by the double journey transmission time. When it does cause difficulty *echo suppressors* are connected at each end of the circuit. These are *voice operated* in that they detect a voice signal in one path of the 4-wire circuit and automatically insert attenuation in the opposite direction to reduce the echo level.

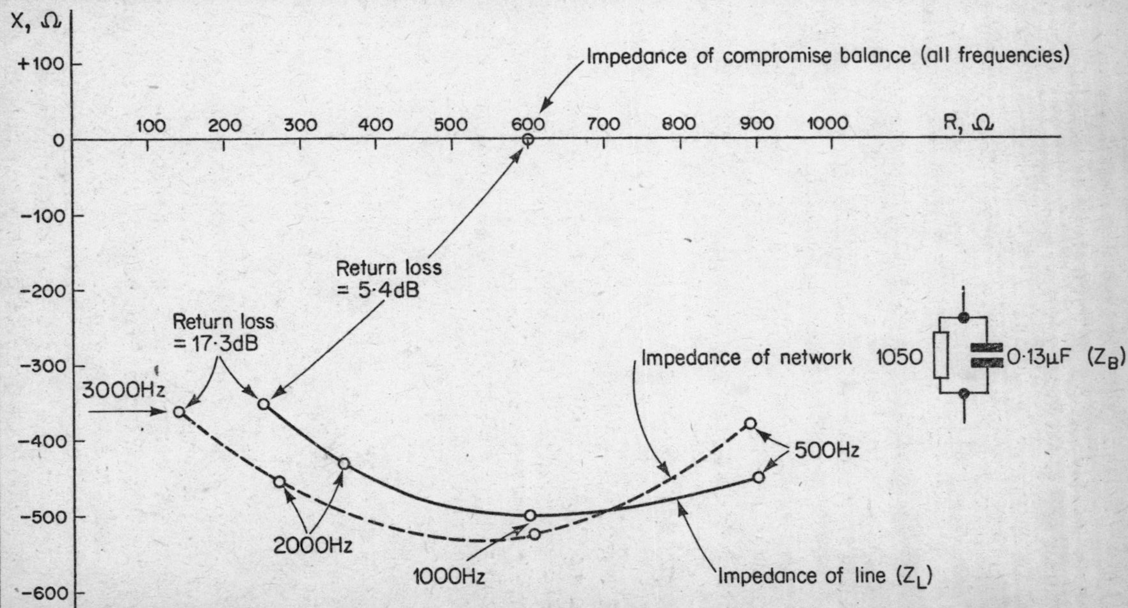

Fig. 4.6 Reduction of return loss

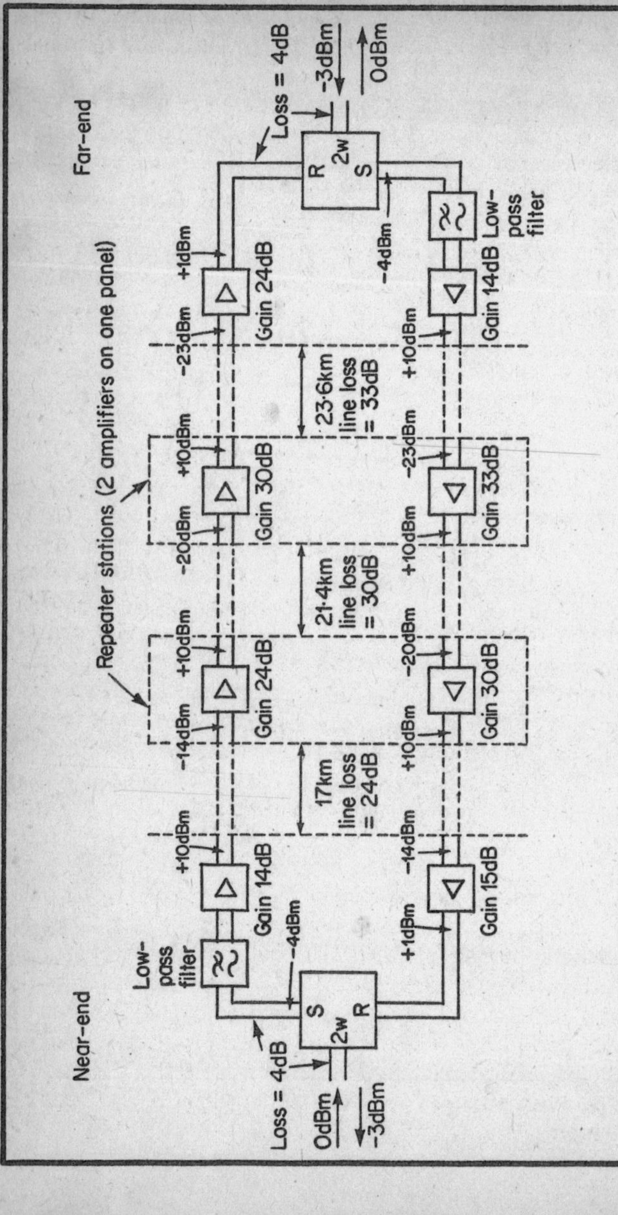

Fig. 4.7 Four-wire audio circuit

For stability circuits are usually adjusted for a small overall loss (say 3dB) in both directions between the 2-wire terminals.

Fig. 4.7 shows a typical circuit connecting two points 62 km apart over underground copper conductors some 0.8mm diameter and having an attenuation of 1.4dB/km. The circuit is arranged to conform with the location of existing *repeater stations* (i.e. buildings containing the necessary amplifiers) as shown. A test signal (usually a sine wave at 800, 1000 or 1600 Hz at a level of 0dBm) is applied to each 2-wire in turn to enable the circuit to be *lined up* by adjustment of each amplifier gain for the desired output level. A low-pass filter is shown in the send path at each end to avoid instability at frequencies above the normal range as explained in the previous section.

The question immediately arises as to why the amplifiers must be spread over the route, why not more conveniently use a single amplifier either at the sending-end or at the receiving-end. Considering the sending-end first, the amplifiers shown as having gains of 14dB would instead need gains of 92dB, (14dB plus the total gain of the amplifiers dispensed with). 92dB corresponds to a power ratio of over 10^9 which would raise 1mW to 1 megawatt (MW), an output more in keeping with that of a power station! On the other hand, with all amplification at the receiving end, the signal level would fall to –91 dBm before amplification, so far below the noise level as to be completely lost. Both methods are therefore out of the question.

Fig 4.7 illustrates a typical medium-distance link. Longer ones simply comprise more line sections and amplifiers, so spaced as to maintain a satisfactory signal/noise ratio, the importance of which is evident from Section 1.5.4. A typical line amplifier employs two transistors with direct coupling and overall negative feedback,(3/3.2.7) having a maximum output level of 50 mW (+17 dBm) and a maximum gain of 30dB. The gain is adjusted by insertion of a fixed attenuator (balanced T-pad) at the input.

4.1.2.1 Phantoms

The a.c. signals normally encountered on a pair of wires are said to be *transverse*, they exist *across* the pair as would be shown by a voltmeter or oscilloscope so connected. There is another form of transmission known as *longitudinal* where such a measurement would read zero because the current is flowing along both wires equally. A circuit arranged for transverse working only is shown in Fig. 4.8(i). The transformers match the equipment to the line at each end for optimum power transfer. In Fig. 4.8(ii) the circuit is developed for both transverse and longitudinal operation by simply centre-tapping the line winding of each transformer. Taking current flow from terminal 1 at the near-end to terminal 1 at the far-end for example, it divides equally at the near-end transformer with no magnetic effect because the two currents flow in opposite directions in the winding. At any point along the line the two equal currents have no pd between them. At the far-end the currents similarly have no effect in the transformer. Fig. 4.8(ii) therefore demonstrates how a 4-wire circuit can economically be adapted to provide an additional 2-wire circuit of half the normal line resistance, the two circuits being electrically separate. The method is known as *phantom* working.

Amplifiers en route are bypassed by a similar technique as shown in Fig. 4.8(iii) so that in 4-wire amplified telephony circuits the phantom can carry signalling currents from end to end for setting up and controlling a call. The manner in which a terminating set is bypassed is evident from Fig. 4.3(iii), the terminals marked SIG being connected to centre-taps on the send and receive line transformer windings.

4.2 HIGH FREQUENCY SYSTEMS

The audio-frequency systems discussed above are so called because they carry such frequencies directly, no higher ones are involved. For many reasons which will become evident as we progress, channels also need to work at higher frequencies whereupon the audio or *baseband* range of frequencies is

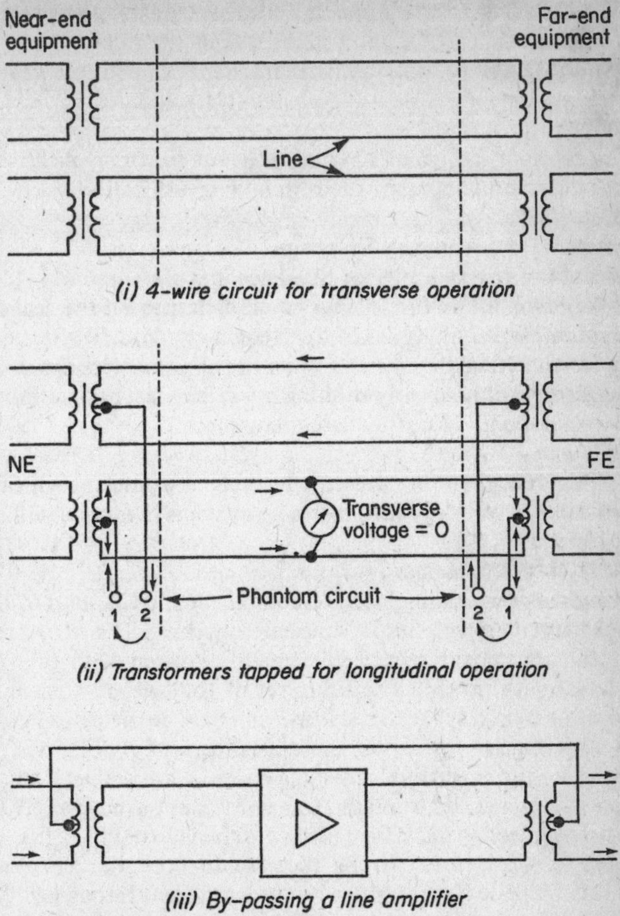

Fig. 4.8 Phantom working

impressed onto a higher one by a process known as *modulation*. An ever-present example is the radio broadcasting service. Audio frequencies cannot be radiated directly (except as sound waves which are soon lost) so a convenient *radio* frequency (say, 150 kHz or more) is chosen and this is made to act as a *carrier* of the speech or music (the baseband) and is then broadcast. Line systems which modulate the information signal onto a higher frequency one in this way are generally known as High Frequency (h.f.) Systems.

The range of frequencies available for use in communication in extensive and is subdivided to show its utilization below.

4.2.1 The Frequency Spectrum

Because of the enormity of the range it is presented in Fig. 4.9 in logarithmic fashion in that equal distances on the scale represent equal multiples in the frequency range. The dotted lines separating the various operational characteristics are suggestions only, clearly nothing is so clear cut, nor are those shown the only uses, there are many more.

The figure is therefore presented simply as a guide to which we can refer as we study line and radio systems in more detail.

4.2.2 Line Systems

It is the overwhelming need to combine many channels over a single circuit which has led to carrier working. The extra cost of the equipment required is small compared with that necessary to provide a separate set of four wires for each additional link. What this leads to can best be understood by considering the quite workable arrangement of Fig. 4.10. Consider the single link providing communication between A and B. At each end is a radio transmitter and a receiver. The A transmitter works at a certain frequency to which the receiver at B is tuned. In the opposite direction the A receiver is tuned to the B transmitter but at a different frequency. The baseband signals have been shifted in frequency for transmission over a radio path. By duplicating the whole system and choosing different radio-path frequencies to avoid interference a second exclusive link can be set up between C and D. The radio link now contains 4 separate channels and itself is called a *broadband* channel, its total bandwidth is at least four times that of a single baseband channel.

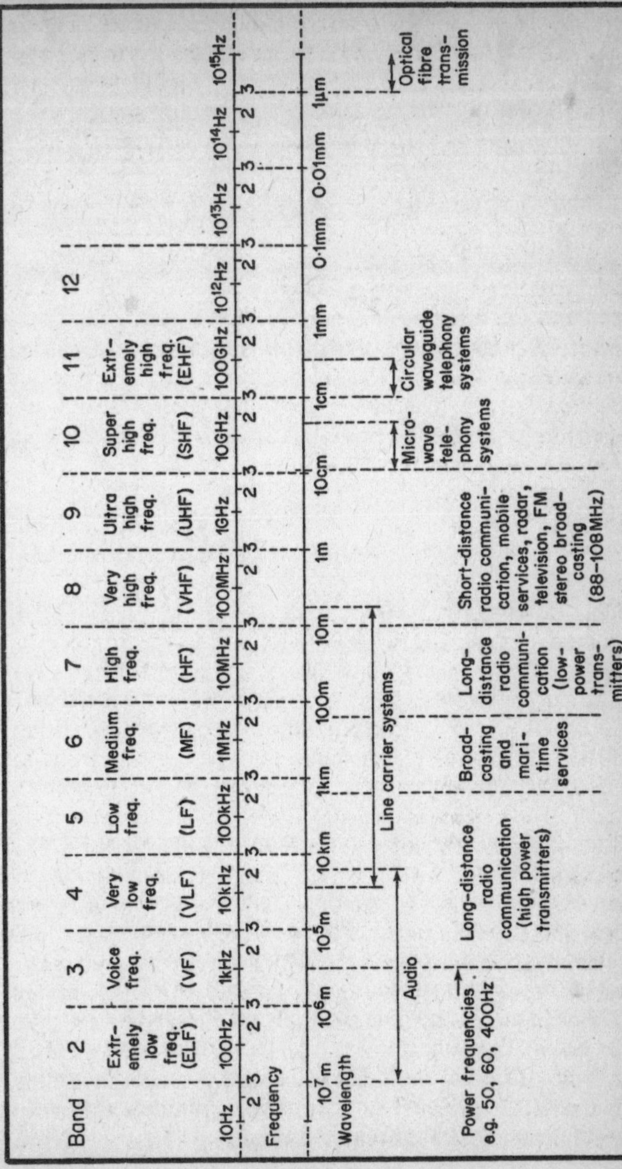

Fig. 4.9 Communication frequency spectrum

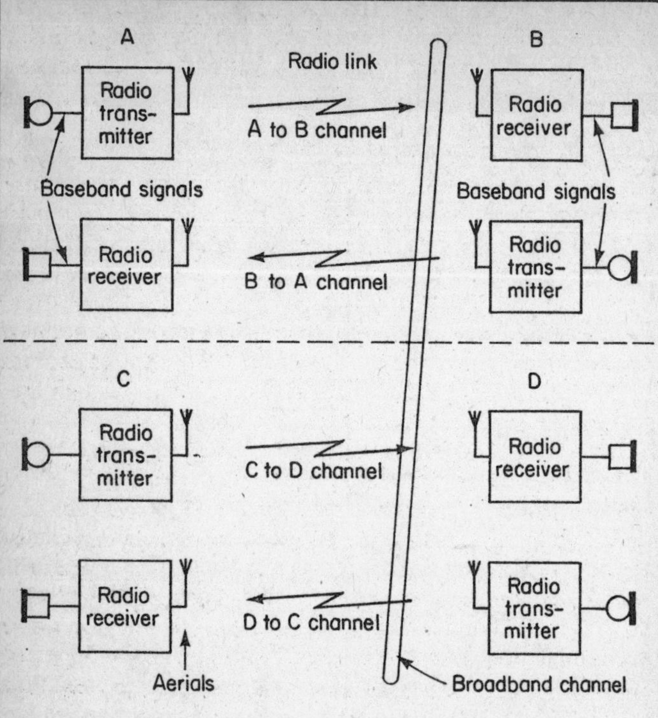

Fig. 4.10 Radio links

With a cable link the same system applies and by using a separate pair of wires in each direction, the A to B and C to D channels are provided on one pair (the *transmit* or *go* pair from A and C) while the B to A and D to C channels are on a different pair (the *receive* or *return* pair to A and C). The bandwidth required on each pair is about twice that of the baseband signal. With wires the transmit and receive frequency bands can be the same and are private, with radio all channels must be on different frequencies because the path is common, not only to the system under consideration but to others. Nevertheless radio links are used in this way.

What we have examined is basically a carrier system and the technique of combining several channels over one path is known as *multiplexing*. This is brought to life by considering practical systems.

4.2.2.1 *The Basic Multiplex Group*

Fig. 4.11 shows the elements of a standard 12-channel *translating equipment* which forms the basic unit of many line systems. The 12 audio channels so combined are said to form a *group* in a *frequency-division multiplex system*. Before looking at the equipment in detail we must bring forward one of the fundamental concepts developed in Chapter 5 on *amplitude modulation* which is the type employed. We must be content here with knowing what modulators and demodulators do rather than *how* they do it.

If a low frequency signal, f_m is mixed in a certain way with a higher frequency, f_c, then three separate frequencies are produced, $(f_c - f_m)$, f_c and $(f_c + f_m)$, the new frequencies being the difference and the sum of the components, they are called *side frequencies* and we see that f_m has been *translated* into two higher values. Now if f_m represents a *band* of frequencies, each single frequency within the band is similarly affected so that the whole band is translated, the result of modulation being two *sidebands* instead of two side frequencies. The two sidebands are distinguished by being known as the *lower sideband* (l.s.b.) and the *upper sideband* (u.s.b.). The principles may be more evident if we consider the speech band generally catered for by such systems, i.e. 0.3 → 3.4 kHz (a slightly smaller band, 0.2 → 3.05 kHz may be used on expensive circuits). Suppose the *carrier* frequency f_c to be 108 kHz, then the three modulation products are (108 kHz minus each frequency in the audio band), 108 kHz and (108 kHz plus each frequency in the audio band), or in figures [108 – (0.3 to 3.4)] kHz, 108 kHz and [108 + (0.3 to 3.4)] kHz.

The type of modulator used suppresses the single frequency f_c and a band pass filter can remove the upper sideband, so we are left with (108 – 0.3) to (108 – 3.4) kHz i.e. 107.7 to 104.6 kHz which at first seems the wrong way round but on

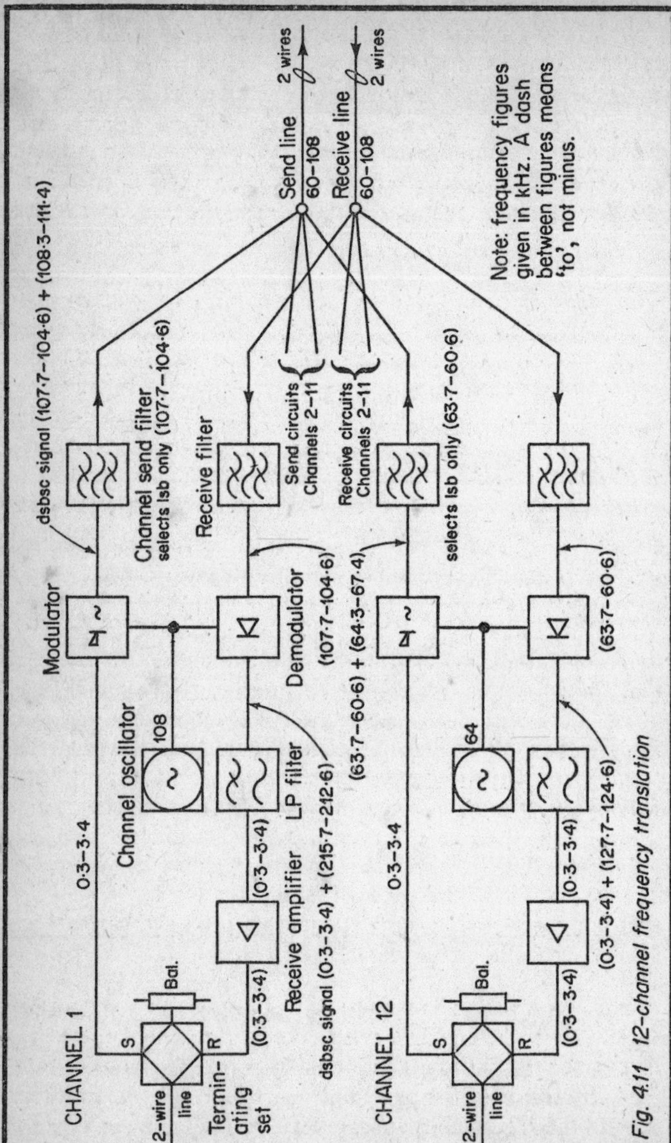

Fig. 4.11 12-channel frequency translation

second thoughts this is how all lower sidebands must work out. Anyway it does not really matter which way round we put the numbers, it is still the same band of frequencies. The audio or baseband is translated into a different frequency band altogether but it contains the same information, none is lost. In demodulation the same process restores the baseband signal for [108 – (107.7 to 104.6)] kHz gives 0.3 to 3.4 kHz. The whole procedure is known as *single sideband suppressed carrier* (s.s.b.s.c.).

This is all we need to know for a general appreciation of Fig. 4.11. The equipment is similar at both ends of the system. Each channel uses a terminating set (Section 4.1.1.1) to separate the sending and receiving paths. On the send side a modulator produces the double-sideband suppressed-carrier signal which is changed to s.s.b.s.c. by the band-pass filter (channel send filter) which follows. This is the band of frequencies sent to line. Channels 1 and 12 only are shown in the figure, each having its own individual frequency slot in the whole band, e.g:

Channel No.	Channel Oscillator Frequency, kHz	Channel Filter Pass Band, kHz
1	108	(104.6 – 107.7)
2	104	(100.6 – 103.7)
¦	¦	¦
11	68	(64.6 – 67.7)
12	64	(60.6 – 63.7)

There is a small gap between adjacent channel pass bands e.g. 103.7 to 104.6 (0.9) kHz between Channels 2 and 1. This allows for the fact that band-pass filters do not have infinite cut-off outside of the pass band (Section 3.3.3). Therefore to obtain the required separation between channels (any overlap gives rise to *overhearing* or *crosstalk* between adjacent channels), a frequency spacing of 4 kHz is used, as is evident from the channel oscillator frequencies. The whole group therefore occupies the band 60.6 to 107.7 kHz, actually made up to 60 → 108 kHz because groups may also be assembled by further translation into larger groups.

The complete band 60 → 108 kHz is transmitted over the send line to the distant end via h.f. amplifiers as necessary to maintain the signal level and ensure a sufficiently good signal/noise ratio. The same range of signal frequencies appears on the receive line and each receive filter accepts its own band of frequencies from the line and rejects all others. The accepted band is passed to the demodulator from which a low-pass filter selects the lower sideband only, this being the baseband (audio) signal. A receiving amplifier raises the audio signal level for transmission via the terminating set to the telephone or equipment at the end of the 2-wire line.

4.2.2.2 Higher-Capacity Systems

By using the 60 → 108 kHz group band to modulate an even higher frequency carrier, separate groups can be assembled to form larger ones in the same way that 12 channels are each translated to form a group. The simplest is the formation of a 24-channel system by adding a basic (60 → 108 kHz) group to a second one which has modulated a 120 kHz carrier frequency and therefore translated its range (again selecting the lower sideband) to (120 – 108) to (120 – 60) kHz = 12 → 60 kHz. The translated and unchanged groups together accommodate 24 channels over the range 12 → 108 kHz.

Similarly 5 basic groups can be so translated as to completely occupy the band 312 → 552 kHz, forming a *basic supergroup* as follows:

Group	Freq. range kHz	Carrier, f_c, kHz	Translated frequency ranges, kHz	Total band kHz
1	60 – 108	420	312 – 360	312
2	60 – 108	468	360 – 408	↓
3	60 – 108	516	408 – 456	
4	60 – 108	564	456 – 504	
5	60 – 108	612	504 – 552	552

The process can be further extended, basic supergroups are assembled as *supergroups* and these as *hypergroups*, 12 of which form a system of 60 MHz carrying as many as 10,800 telephony channels. The build-up of high capacity systems may vary in different countries, but the principles are unaltered. The line is coaxial for both underground and submarine routes (Section 1.6.2.3).

We must dispel the impression that carrier systems are invariably used to carry large numbers of channels, in some circumstances they are economic with no more than a few. Overhead open-wire systems in sparsely populated areas may use a few up to a dozen or so channels usually with a different band of frequencies for the send and receive directions so that the whole system can run over two wires instead of four. A large number of channels is not usually a practical proposition over an overhead line because all circuits are lost at once on a line failure, a broken overhead wire is certainly not an uncommon event.

Compared with that for speech, the information rate for both computer and television signals is very high. For a given signal/noise ratio Shannon's formula indicates that such signals need a high bandwidth. H.F. systems cater for these by allocating a block of bandwidth as required, for example, for high-speed data transmission a basic group of 48 kHz may be used as a whole instead of splitting it up for 12 telephony channels. Very much more is required for television, for colour signals require some 5 – 6 MHz. For fairly short distances, because of the wide band needed, t.v. signals are often transmitted over a single coaxial cable.

4.2.2.3 Submarine Systems

It is the development of submerged repeaters to allow the use of multichannel carrier systems which has continually reduced the cost of submarine cable systems since their inception. Compared with a land-based h.f. system the submarine one has two noteworthy differences:

(i) power for the repeaters cannot be fed in at any point in the sea, hence it must be supplied from the land-based ends
(ii) by using different frequency bands for transmit and receive, a single modern cable is capable of carrying both directions of transmission.

Coaxial cables of some one inch diameter are used with repeaters housed in high-tensile steel tubular cases capable of withstanding the very high pressure at the bottom of an ocean. Their size is such that they appear as no more than a large bulge in the cable itself.

System capacities vary greatly depending on length and type of cable. The greater the number of channels, the wider the frequency band required and the greater the highest frequency loss, consequently with lower repeater spacing. System frequencies range up to 40 – 50 MHz, catering for 5000 channels or more, although well over 10,000 seem possible. System capacities using optical fibres are mentioned at the end of Chapter 7.

Two examples using the same cable are:

No. of channels	Highest frequency	Approx repeater spacing
80	608 kHz	29nm* (about 54 km)
360	2964 kHz	9.5nm (about 17.5km)

(*nm = nautical mile = 1.15 miles = 1.853 km)

showing once again that as the signal becomes attenuated and the signal/noise ratio worsens, amplification is required more frequently.

Power is fed along the centre conductor of the cable with the broadband signal and separated out at each repeater. The amplifier may need 50 or more volts, the result of many in series being that a high-voltage d.c. supply is required. To reduce the voltage applied at the end of the cable, the supply is split into two, half at each end as in Fig. 4.12(i) which shows that for a typical modern system using a 5500V supply for 96 repeaters at 50V, the maximum voltage applied to the cable is 2750. The path of the current (390mA) is in series through the repeaters and returns via the sea which provides an excellent "earth" connexion of very low resistance. There is of course, a loss of voltage due to the resistance of the cable centre conductor. Some Ohm's Law calculations show the distribution of resistance:

Total power circuit resistance $= \dfrac{5500}{0.39} = 14102\Omega$

Total resistance of repeaters $= \dfrac{50}{0.39} \times 96 = 12308\Omega$

Total resistance of Cable (+ Sea) $= 14102 - 12308 = 1794\Omega$

$\therefore$ Resistance of Single Cable Section $= \dfrac{1794}{96 + 1} \approx 18.5\Omega$
(neglecting sea resistance)

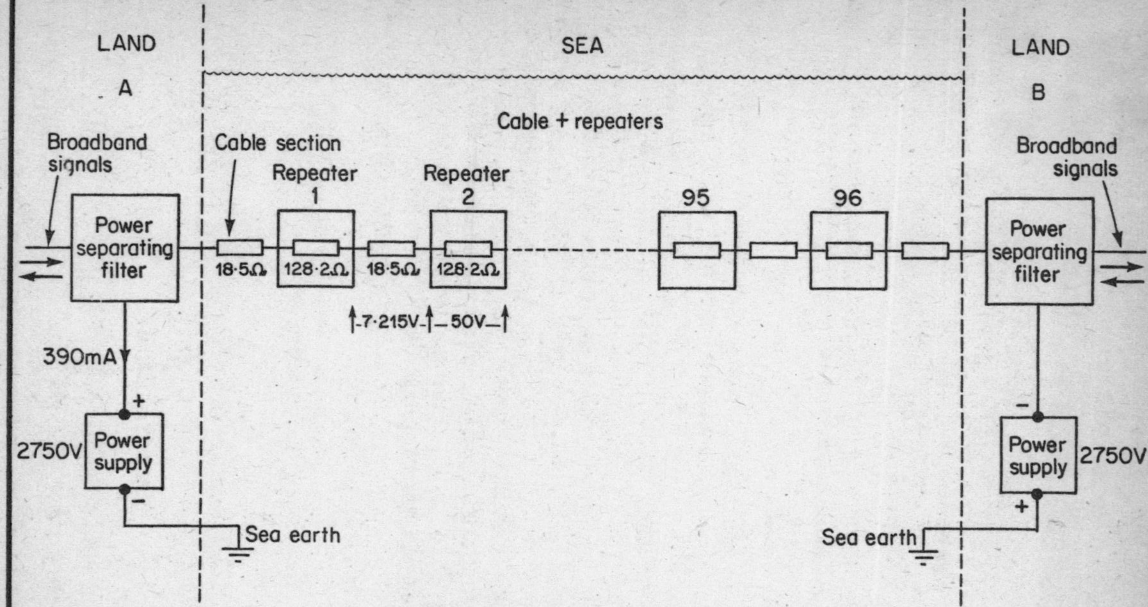

(i) Power-feeding arrangement

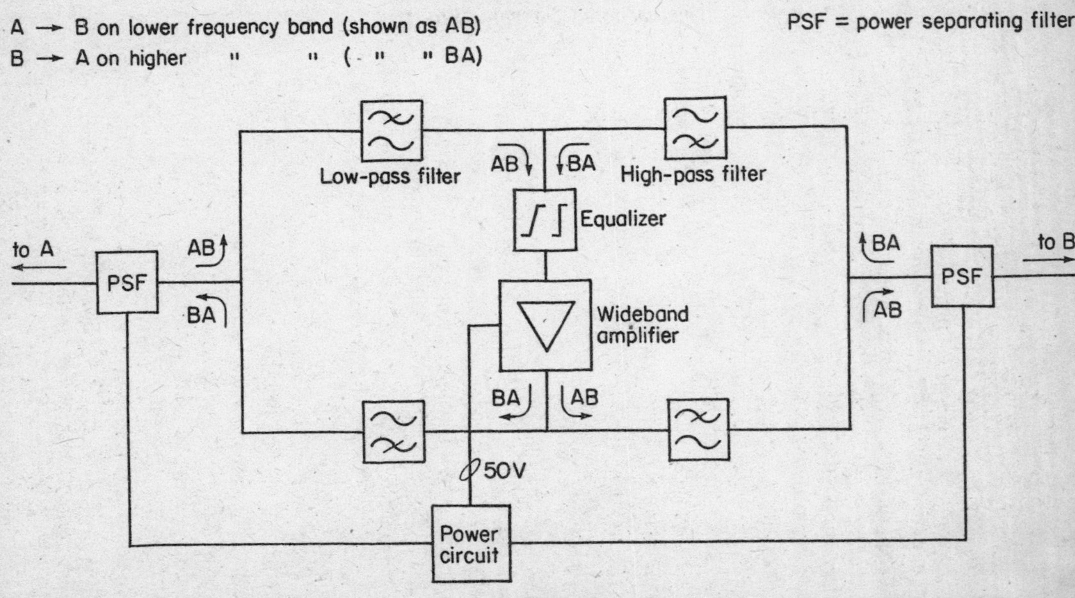

(ii) Repeater

Fig. 4.12 Typical submarine system

from which voltage drops can be calculated as shown in the figure.

Fig. 4.12(ii) shows the main features of a typical repeater. By a system of high and low-pass filters the two frequency bands are separated and directed through the single wideband amplifier (usually 3 stages with negative feedback(3/3.2.7) and a gain of some 40 – 60dB). The two broadbands have sufficient frequency separation between them to provide good filtering, for example A to B might be carried by 312 – 1428 kHz with B to A by 1848 – 2964 kHz. An *equalizer* is necessary to compensate for the increase in attenuation with frequency of the cable. It does this by adding attenuation at the lower frequencies but not at the higher so as to produce an overall attenuation which is the same at all frequencies. (Equalizers are also used with h.f. land systems but were not introduced earlier so as to avoid overcomplication). The power circuit bypasses the filters and is connected to the amplifier as shown.

4.2.3 Radio Systems

Frequency translation in h.f. line systems applies equally to radio systems which transmit the modulated carrier as an *electromagnetic wave* through the atmosphere or space. How this is done we look at in greater depth in Chapter 6. In a way, radio signals all use the same transmission medium so control must be exercised in the allocation of frequency bands to various countries. Over the range (as shown in Fig. 4.9) from say, 10 kHz to about 20 GHz propagation conditions vary widely and by choice of suitable carrier frequency so that the wave is not effective outside a given area or by the use of directional aerials which focus most of the radiated energy on one point only, most services are provided without overlap. An example is given by the broadcasting band shown in Fig. 4.9, at these frequencies the radio signal suffers relatively high attenuation with distance, thus the same ones can be used in many countries without interference.

A signal of any frequency has a corresponding wavelength(2/1.2.6) and it is perhaps confusing that radio transmissions may be designated by either, we talk of long, medium and short-*wave* transmissions but higher up the scale change to frequencies especially throughout the MHz range. Ultimately for microwave and optical systems "wavelength" comes back into its own because of its relationship with dimensions of aerials and transmission guides (this will become evident later). The formulae for conversion between frequency (f) and wavelength (λ) are simple fractions i.e.

$$\lambda = \frac{v}{f}\text{metres} \quad \text{or} \quad f = \frac{v}{\lambda}\text{Hz}$$

where v is the velocity of radio waves, 3×10^8 m/s (note that the velocity of signals over lines is not the same as we will see in Chapter 6). For quick conversions and as a check on calculations, Appendix 1 is helpful, giving reasonably accurate answers without too much manipulation of decimal points or powers of 10. Interpolation can improve the accuracy where necessary, for example, 197 kHz has no exactly corresponding figures in Column 1 but since it is about half-way between 195 and 200, a Column 2 figure might be guessed as 15.20 (about half-way between 15.00 and 15.38). This gives an answer of 1520m. The calculated value is 1523m hence the error is small. Fig. 4.9 is also useful as a rough guide.

4.2.3.1 Broadcasting

Radio broadcasting to the general public is a one-way system employed universally. For speech and music transmission two different modulation principles are used (i) amplitude modulation (a.m., as used for h.f. line systems – Section 4.2) and (ii) *frequency modulation* (f.m., to be discussed further in Chapter 5) which is a different technique by which a signal is impressed on a carrier wave. FM has certain advantages, it is transmitted using higher carrier frequencies than for a.m. and carries a wider baseband signal for higher fidelity reproduction. Thus whereas a.m. transmissions occur mainly over the LF and MF bands from about 140 kHz to 1.8 MHz, f.m. is carried in the VHF band between 70 and 110 MHz. Stereophonic transmissions which embrace two separate channels are also broadcast in this range. At still higher frequencies is television broadcasting, mainly over two ranges, about 170 – 220 MHz and 470 – 900 MHz.

We are already persuaded that a system such as television which needs a higher information rate than for speech or music must have a greater transmission bandwidth. When used as a baseband signal to modulate a carrier it also needs a higher carrier frequency. That any carrier frequency must always be considerably greater than the maximum baseband frequency begins to make some sense for by taking the reasoning to the extreme, a 6 MHz bandwidth television signal cannot modulate a 1 MHz carrier because the result would be a sideband extending 5 MHz below zero frequency, an impossible situation. Yet such a carrier is adequate for audio signals in the medium waveband.

4.2.3.2 Microwave Systems (Terrestrial)

Frequencies higher than 1 GHz (10^9 Hz) are also referred to as *microwaves* for at this point λ is a mere 30 cm compared with several hundreds of metres for the longer wave broadcast bands. Microwave transmission systems use wavelengths down to about 2.5 cm as shown in Fig. 4.9. Such systems have the same basic format as h.f. systems, i.e. bothway frequency-division multiplex. Terrestrial (earth-bound) systems are known as *radio relay* systems and these are complementary to h.f. line ones, doing the same job in a different way. There are conditions however where there is no alternative to a radio relay system, such as over mountainous or rocky territory where cable-laying is impracticable. Repeater stations are employed en route and system lengths of up to 6000 km are possible.

In *satellite* systems the signals are transmitted to an Earth satellite, amplified and retransmitted to the receiving station in one *hop*. There can only be one repeater station, in the satellite. The two methods of microwave transmission are considered separately although we shall find that fundamentally they have much in common.

As with line h.f. systems, the radio relay may be an essential part of the trunk network (the major links between cities and towns of a country). By working at such high frequencies, wide frequency bands, known as *broadbands* may be carried so that it is possible for a single *r.f.* channel (the basic radio relay unit) to accommodate more than 2000 telephony circuits, usually built up as a standard f.d.m. block (Section 4.2.2.1). Several r.f. channels may be combined to give a total system capacity of over 20,000 circuits. Needless to say, with such capacities, reliability is of vital concern, sudden failure of a system could cut off thousands of telephone users at once, thus the technique of immediate change-over (within a few μs) to r.f. channels standing idle for this eventuality is often used.

We have discussed amplitude modulation only sufficiently to understand what it does rather than how it is done and now do the same for frequency modulation. Recalling that to impress baseband signal information onto a carrier wave, the latter must change in some way in accordance with the signal variations, f.m. is simply a technique in which the carrier frequency is varied about its mean value in accordance with the amplitude of the modulating wave. The amplitude of the frequency-modulated wave does not vary and therefore carries no information. Difficult to appreciate though this may be, it is suggested that the reader should feel no anxiety at this stage, we are trying to get a feeling for the whole process before we discuss technical details in Chapter 5.

Because the amplitude of the f.m. wave does not vary, any noise added which increases the amplitude can be cut off by a *limiter* at the receiver with no loss of information, this being entirely contained in the rate and degree of frequency change. Hence the received signal suffers less from radio path noise which is generally more troublesome than with cables. The cost of the improved signal/noise ratio is as might be expected, in the requirement of a larger bandwidth for f.m. Sufficient bandwidth for multiplex systems is available at frequencies in the gigahertz range.

Just as a motor car headlamp concentrates the light from the bulb into a more or less parallel beam, so also can reflectors be designed for use with radio waves. Circular reflectors are usually used and the diameter must be several times (10 or

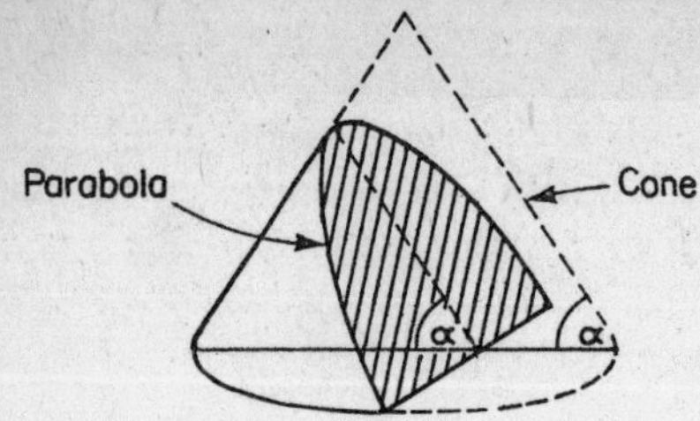

(i) Cone sliced parallel to side

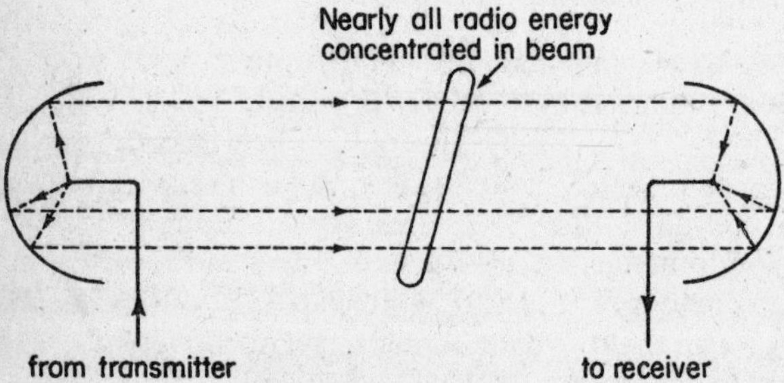

(iii) Microwave aerial link

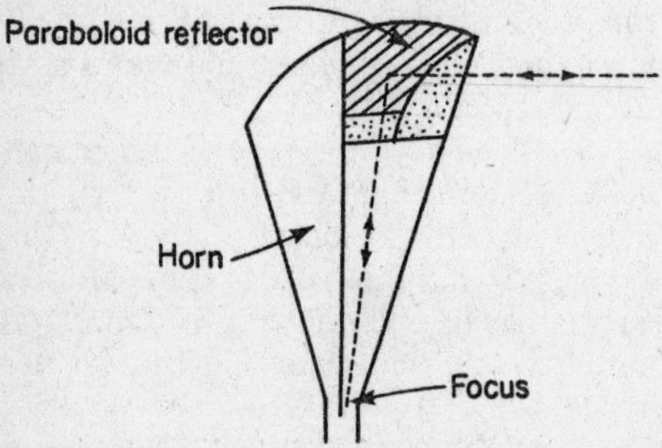

(v) Horn-type aerial

Fig. 4.13 Microwave aerials

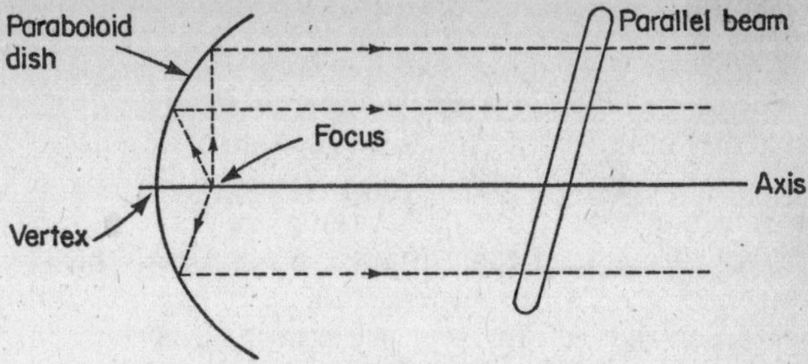

(ii) Wave generated at focus of transmitting aerial

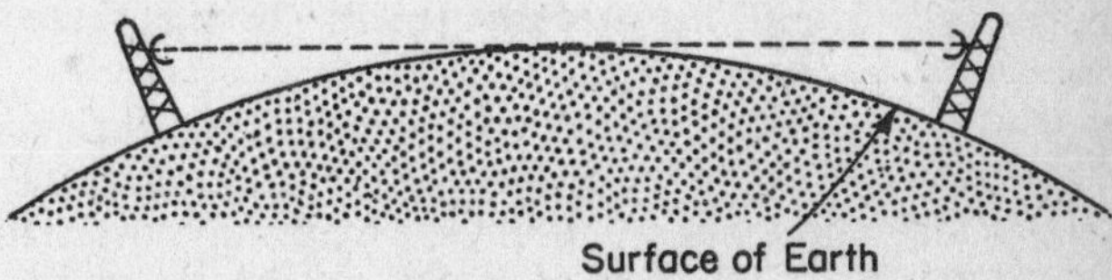

(iv) Line-of-sight path

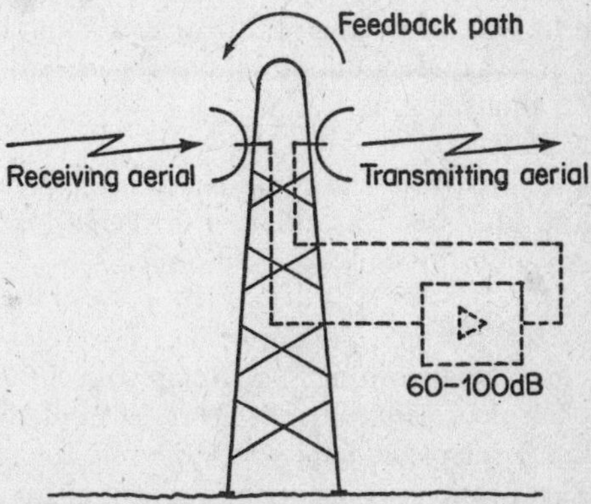

(vi) Disposition of aerials on repeater station mast

more) the wavelength hence the technique is only usable with short wavelengths. As an example, at 3 GHz, λ = 10cm, requiring at least a 1m diameter reflector (we begin to see why we talk in terms of λ rather than f at these frequencies). The shape of the reflector is *parabolic* because a parabola has the capability of reflecting radio waves or light from a point known as its *focus* into a parallel beam. It also works in the reverse direction, that is, a parallel beam when received along the axis is reflected and concentrated on the focus. This is illustrated in Fig. 4.13. We get some idea of the shape of a parabola from (i), it is formed when a cone is sliced parallel to its side as shown. If a radio signal of suitable wavelength is generated at the focus it is directed by a parabolic reflector (which may be of solid metal or wire mesh) into a parallel beam as shown in (ii). As a receiving aerial a parallel beam is reflected to the focus at which is installed a collector. Thus we can have point-to-point propagation as in (iii), very little radio wave energy spreads out to be lost and the system does not broadcast, an important constraint when private conversations are being carried. The axes of the two aerials must be accurately in line. Now comes the problem, the curvature of the earth interferes with the beam as distance between the aerials increases (iv). This can be offset to a certain degree by the use of high aerial towers (100m or so and in cities with extra height to avoid all buildings) but generally distances do not exceed about 60 km.

For those readers who wish to understand the parabola a little better some additional information involving practice with graphs is given in Appendix 2 (Section A2.3).

There are occasions when microwave aerials are encountered which seem to have nothing in common with the circular dish, an example is sketched in Fig. 4.13 (v). These do in fact work on the same principle and are *horn-type* aerials containing a paraboloid reflecting surface with the focus at the centre of the narrow end. One of the advantages of this type is better shielding at the sides which minimizes interference between adjacent aerials. The importance of this becomes evident later in this section.

Showing diagrammatically the elements of a microwave system gives us the opportunity of getting all we have learned about multiplex systems, both line and radio, into some overall perspective. Thus instead of illustrating a microwave link only in Fig. 4.14 the translating equipment (but for a mere 120 circuits as an example) is also shown. From this figure we see how a telephone call progresses over a trunk system. World-wide calls follow the same basic principles, any number of 4-wire amplified, h.f. land, or submarine cable links or microwave links may be employed in any order. International control ensures that signals from one country have the correct characteristics to suit the links of another. On many calls a satellite or digital link is included, these are discussed later in this chapter.

From Fig. 4.14 can be traced the communication link between two telephone users. A second complete channel through the same links but in the direction far-end to near-end completes the overall link for two-way conversation.

(i) the telephone is connected to its *local* exchange ("central office" in the USA) and is switched from there over a 2-wire or 4-wire *junction* line (toll-collecting trunk) to a *trunk* (toll) exchange. It is usually from the latter that multiplex systems are economical. The audio frequencies are translated to a channel in the range 60 – 108 kHz by a group translating equipment (g.t.e.), the figure shows a connexion to Channel 8 of GTE 1. All five g.t.e.'s are further combined by frequency translation into the range 312 – 552 kHz (Section 4.2.2.2) and again combined with the output from a second set of five g.t.e.'s to produce a *broadband* 312 – 792 kHz. In this form all 120 channels are transmitted to the near-end microwave terminal (which could be in the same building) where they are amplified if required and then frequency modulated onto a 70 MHz carrier (modulation and amplification are easier at this lower frequency than at the microwave frequency). An amplifier follows, then a *mixer* fed by a microwave oscillator (at 4.07 GHz in this example). The mixer is simply an amplitude modulator used for

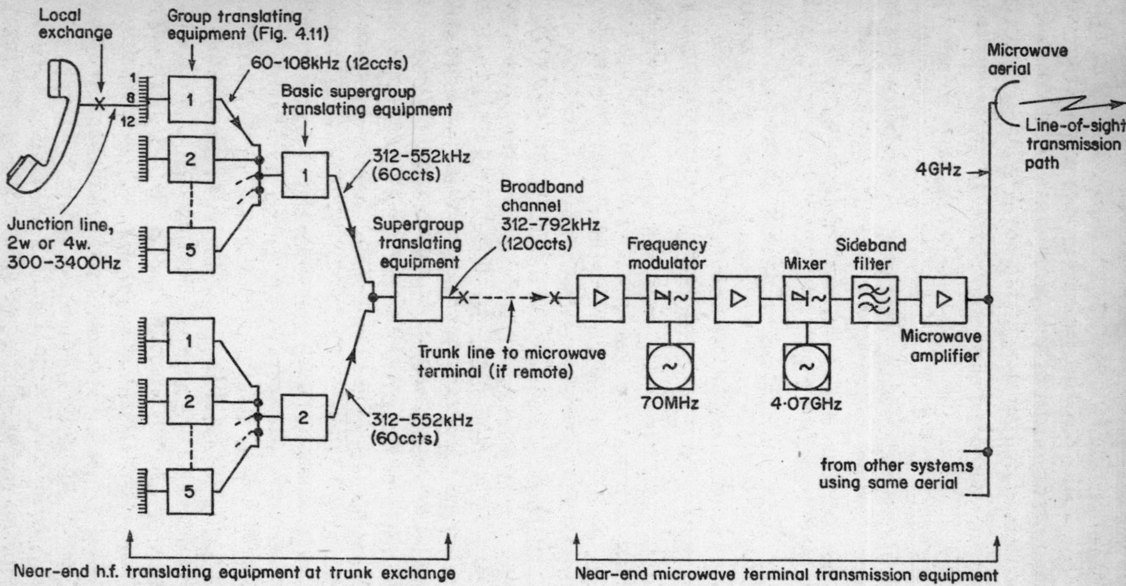

Fig. 4.14 Telephone communication over a microwave link, (i) Near-end transmitting equipment *cont. ...*

frequency translation. Sidebands are produced and a sideband filter selects the desired one, usually the lower, giving 4.07 – 0.07 = 4 GHz. A special type of amplifier provides the necessary aerial power (a few watts) at microwave frequencies, the signals from this are then fed to the aerial.

(ii) every 40 to 60 km a repeater station receives the weak modulated radio signal, in the figure one such station only is shown. The signal is fed from the aerial to a bank of band-pass filters (several systems on different frequencies may be handled by the same aerial), our particular frequency band is selected and mixed with, say 4.07 GHz to produce an *intermediate frequency* (i.f.) which can be amplified more easily than the incoming microwave signal. Changing to an i.f. is in fact the *superheterodyne* principle found in most radio receivers (more on this in Section 5.2.5). Two i.f. (70 MHz) amplifiers then follow with a limiter in between. Remember the figures quoted are the various carriers, the actual signal is really a sideband but in fact with frequencies only a very small fraction of 1% different from the carrier.

The amplified 70 MHz signal is again mixed but this time with an oscillator frequency which produces a transmitted frequency slightly different from the received one, in this case 4.2 GHz instead of 4.0 GHz. The overall gain of a repeater station is probably some 60 – 100dB as shown in Fig. 4.13(vi). Now the aerials are very directional and in an ideal situation no energy would be fed backwards from the high output power transmitting aerial. However some finds its way through various paths and reflexions back to the receiving aerial which results in instability if the *loop* gain (transmitting aerial → receiving aerial → amplifier → transmitting aerial) exceeds unity (0dB) at any frequency.[(3/3.3)] Thus an attenuation of some 100dB between the two aerials may be required; not always easy to obtain. An additional way of minimizing feedback is to transmit on a slightly different frequency. Thus the figure shows a mixing oscillator at 4.27 GHz with a sideband filter to select 4.2 GHz which is power amplified and transmitted onwards. There may be many such repeater stations en route and at each successive one

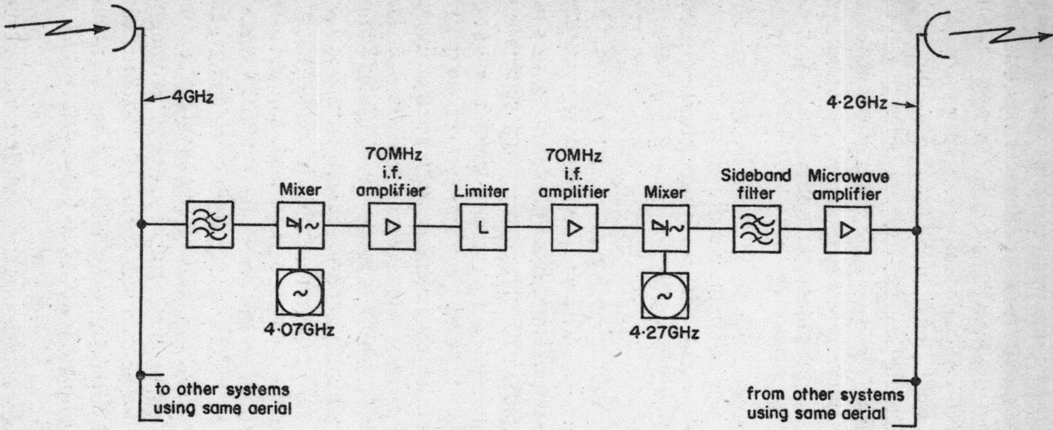

Fig. 4.14 cont. (ii) Repeater station

cont. ...

the frequency shift is reversed, i.e. Repeater Station 2 receives on 4.2 GHz and retransmits on 4.0 GHz.

(iii) ultimately the far-end microwave terminal receiving aerial picks up the signal from the last repeater station (but first in the return direction) and it is selected by a band-pass filter then mixed to generate a 70 MHz i.f. for amplification and limiting. It is then demodulated to regain the broad-band. Amplification follows and the original range of frequencies, 312 – 792 kHz is divided down to the group demodulating equipments, thence to the appropriate base-band channels and onwards locally to the receiving telephone.

4.2.3.3 Microwave Systems (Satellite)

The restriction on distance between microwave terrestrial repeater stations arises mainly from two factors, the effect of the curvature of the earth [Fig. 4.13(iv)] and the attenuation of the wave through the absorption of energy by water and rain in the atmosphere. Satellite communication eliminates the first and greatly reduces the second for the wave travels mostly through space. Satellites carry a high proportion of the World's international traffic and the geo-stationary type (geo, from Greek, earth) is now used almost exclusively. Their advantage over other types is that their position in space relative to any point on earth is fixed, hence complicated aerial "tracking" arrangements are not required. Fig. 4.15(i) illustrates the principle. The satellite is in a circular orbit (1/A8) the plane of which cuts the Earth at the equator, its speed is such that it completes one orbit in exactly the same time as the Earth takes to rotate once, i.e. 24 hours. If we imagine an observer on Earth looking at the satellite (it is much too far away in fact), then as the latter moves through so many degrees of its circular path, the Earth has also turned by the same angle so the observer still looks in the same direction to "see" it. Hence although the satellite is actually moving fast, to the observer on Earth it appears to be stationary.

Fig. 4.15(ii) shows some of the dimensions and distances. They are approximate but serve to show the basic simplicity of the idea. This is a plan of the arrangement, we imagine that

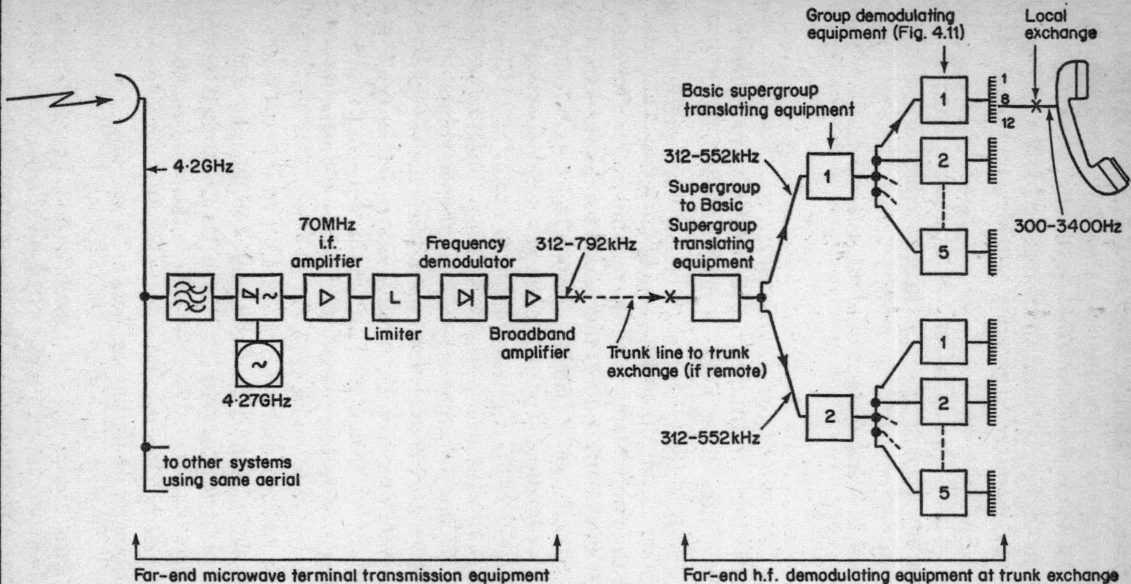

Fig. 4.14 cont. (iii) Far-end receiving equipment

we are up above, looking down on the North Pole. The satellite speed is such that it needs to be some 36,000 km away from the Earth to obtain a 24 hour orbit the radius of which is some 36,000 + 6370 = 42,370 km. The total length of orbit is given by the circumference of a circle of this radius, i.e. $2\pi r = 2\pi \times 42370 = 266{,}220$km, therefore the speed of travel of the satellite is

$$\frac{266{,}220}{24} = 11090 \text{ km/h}.$$

These figures are for demonstration only, the exact ones at any time are determined by *command* signals from the ground which position the satellite exactly.

It would at first appear that two satellites diametrically opposite could provide coverage of the whole earth but it is evident from Fig. 4.15(ii) that near the boundary the transmission path is almost parallel to the ground and therefore via an excessive amount of atmosphere. Hence three satellites (known as *Intelsats*) are equally spaced on a common orbit and provide full earth coverage, obviously with some overlap. They are positioned above the equator at the Indian, Atlantic and Pacific Oceans. The launching of a satellite and positioning in orbit is a highly complex operation while maintaining it in the correct position needs fine adjustment so that not only does its rotation match that of the Earth but also its directional aerials are always aligned with those of the earth-stations. For this a *command receiver* is installed within the satellite for reception of control signals with a telemetry (measurement at a distance) transmitter to advise the controlling station of working voltages, currents etc. The main communications equipment has the same form and function as that in a terrestrial microwave repeater station except that there is one direction only, so needing a directional receiving aerial, filters, mixers, amplifier and limiter, ultimately feeding a directional transmitting aerial. Fig. 4.14(ii) is a reminder of what is included. Frequencies are around 6 GHz earth-to-satellite and 4 GHz satellite-to-earth with bandwidths sufficient for up to 12,000 telephone plus two t.v. channels. The various countries using the international satellite system are each allocated a different frequency band.

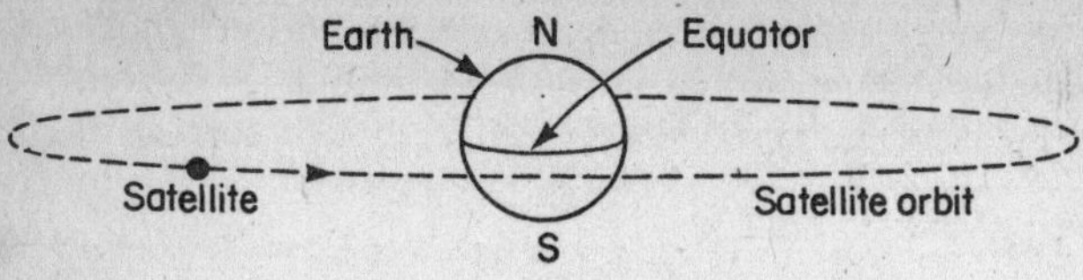

(i) Orbit in equatorial plane

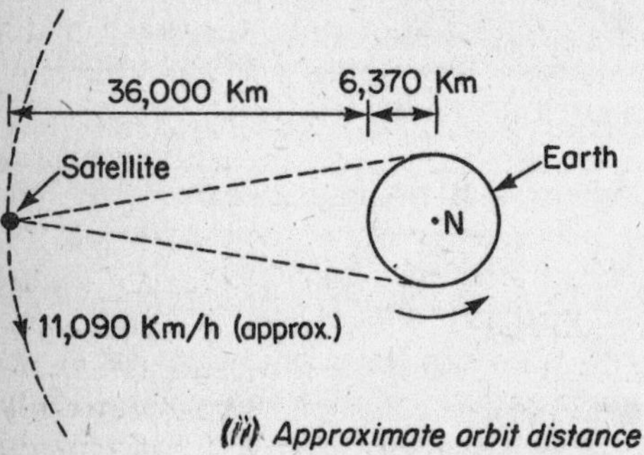

(ii) Approximate orbit distance

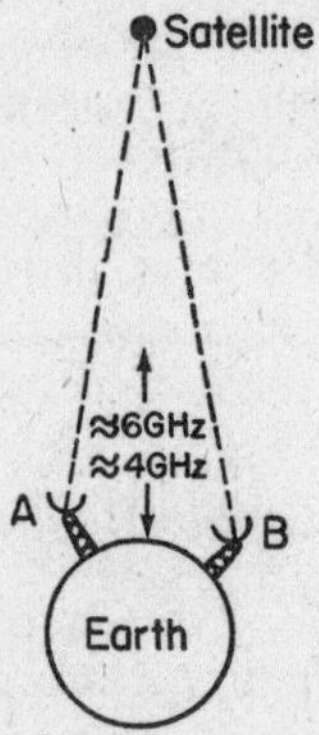

(iii) Transmission path via satellite

Fig. 4.15 Geo-stationary satellites

Power for the equipment is obtained from sunlight by banks of silicon solar cells, each having a d.c. output voltage of about 0.5 at which, with moderate sunlight some one or two hundred milliamperes of current are available. Current satellites are some 7m high and weigh about 1 tonne.

Because of the great distances and therefore attenuations, and the restrictions on availability of power and size of aerials in the satellite, the earth-stations must be very efficient and capable not only of receiving very weak signals but of amplification without engulfing them in noise. We recall that resistance noise varies as the square-root of the temperature,[(3/3.2.2.2)] hence special amplifiers are used, frequently at extremely low temperatures, only a few degrees above absolute zero.

The axes of earth-station aerials are lined up accurately with those of the satellite and a typical transmission path between two stations A and B is shown in Fig. 4.15(iii). The aerial follows the basic principle of parabolic reflexion but usually with an additional feature to further improve performance. The geometry of such a *Cassegrain* aerial system (after N. Cassegrain, the French designer of early reflecting telescopes) is shown in Fig. 4.16. It avoids the difficulty caused by having the feed at the focus and the loss and interference due to forward radiation. A *hyperbolic* reflector is situated between the main dish and its focus. The *hyperbola* is also a conic section as in Fig. 4.13(i) but the section is made at an angle to the base greater than that of the side. i.e. greater than α as shown for the parabola. Again the aerial transmits a parallel radio beam or equally receives one and directs it to the centre of the main dish. The larger aerials are over 30m in diameter and weigh some 300 tonnes.

An operational difficulty introduced by satellites is of *delay*. Fig. 4.15(ii) shows that the signal must traverse at least 2 x 36,000 km in its journey from talker to listener. It is a radio signal and we know its velocity hence

$$propagation\ time = \frac{2 \times 36{,}000 \times 1000}{3 \times 10^8}\ s$$

$$= 0.24s = 240ms$$

This is the shortest time, that is, for both stations directly below the satellite, on average the figure becomes about 270ms, twice this between speaking and receipt of an answer. This is just tolerable but the delay for two satellite links in tandem is not, it causes confusion in conversation. Hence international calls do not contain more than one satellite *hop*. Echo suppressors (Section 4.1.2) are also necessary.

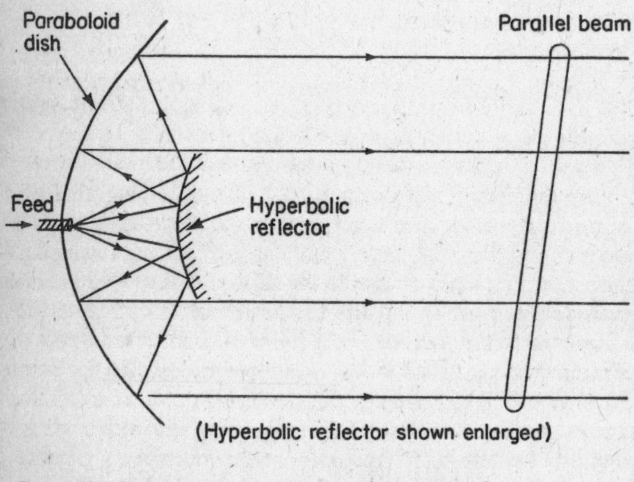

Fig. 4.16 Cassegrain aerial system (transmitting)

4.2.3.4 Mobile Services

Services in which one or both of the participants is mobile can only be provided by radio. Usually the VHF range (Fig. 4.9) is used for which propagation is mainly line-of-sight as with microwaves. However radio waves can be *refracted* so that they are to a certain extent bent round the earth, so partly avoiding the path-length restriction depicted in Fig.4.13(iv).

Refraction is the term used for the change in direction of a wave (or ray of light) in passing from one medium into a second one in which its velocity is different. In this case it arises from the fact that water vapour is present in the atmosphere but its concentration gets less with height, the velocity of the wave becomes lower nearer the earth, consequently the wavefront tilts slightly towards the earth (just as a motor car tends to move from a straight line into a circle if a brake rubs on one wheel). This is obviously an incomplete explanation but to examine in depth the mechanism of wave refraction is neither necessary here nor straightforward without having first studied the radio wave itself. Refraction is explained more fully in Appendix 2, Section A2.2.

Under certain conditions it is possible for the degree of refraction to be sufficient for the wave to follow the earth continually, making possible propagation over very long distances. Full advantage cannot be taken of this however because atmospheric conditions change.

We have seen that a.m. requires less bandwidth than f.m. and is accordingly used for mobile radio because often there is a large demand for channels within a given area, for example, for police, taxis, fire and ambulance services, motoring patrols and private services. When only one carrier frequency is available to a system, interference is avoided between stations by a rigid operating procedure which allows a station to hold the channel until it announces completion ("over" etc). This is necessary because two stations transmitting together, although nominally on the same frequency are almost certain to be slightly different. In the next chapter we see that when two sinewaves which differ slightly in frequency are added together (this is not modulation because neither wave need carry information), one resulting component is the frequency difference between them. For example suppose two transmitters have a nominal frequency of 80 MHz but that one is "off-frequency" by a mere 0.0005%. The two carriers are therefore

(i) 80 MHz and

(ii) $80 + \left(\frac{0.0005}{100} \times 80\right)\text{MHz} = 80.0004\ \text{MHz}.$

The difference between them is 400 Hz and this will appear on the circuit as a continuous whistle (explained further in Section 5.1), making conversation difficult or even impossible. Thus one transmitter only can operate at a time.

For telephone service between the public network and vehicles separate channels are required for all calls in progress at any one time otherwise privacy is lost. The operating method usually adopted is to use one channel exclusively for contacting wanted subscribers or for subscribers to call the operator, i.e. a signalling channel. The operator then allocates free speech channels for the calls.

4.2.3.5 H.F. Services

Except for satellite systems we have not yet discovered how a long-distance point-to-point service is provided. In fact before submarine cables and satellites were developed sufficiently, radio was the only means of providing telephone service between countries and still is in use for long-distance communication with ships and aircraft.

The *ground* wave which is used at the lower end of the frequency spectrum is unreliable for long distances (thousands of km) because it is rapidly attenuated in its passage over the earth, the microwave at the other end of the spectrum expires at even shorter distances. Fortunately in between, that is, at h.f. a natural phenomenon comes to our aid. Well above the Earth, between about 50 and 400 km high, *ionized layers* exist in what is appropriately called the *ionosphere*. Our earlier studies show that when an electron receives sufficient energy it can escape from its orbit leaving the atom positively charged and known as a positive *ion* until it captures an electron to become electrically neutral again. Energy for this to happen is supplied by the ultraviolet rays in sunlight which are more effective at these heights than nearer the ground where more atmosphere has been traversed, thus giving a lower limit to the layer. An upper limit is also easy to visualize because the air is becoming progressively more rarified as space is approached. We can therefore simply imagine a layer one or two hundred kilometres deep with the greatest ionization at the centre, diminishing above and below.

An abbreviated explanation is that free electrons in the ionized layer receive energy from the wave and vibrate. Some of this energy is returned in a different phase, the effect on the wave being to bend it away from regions of high electron density to lower. As the wave penetrates the layer the *refraction* which is proportional to free-electron density increases and pictorially the effect is as shown in Fig. 4.17(i) while (ii) shows how the effect enables radio communication to be possible over long distances. Refraction is considered more fully in Appendix 2.

Because bandwidth is at a premium, amplitude modulation is invariably used often with *independent sidebands* in which each of the two sidebands arises from a different channel, that is, one conversation is carried by the l.s.b. and a different one by the u.s.b.

4.3 DIGITAL SYSTEMS

One may have the feeling that digital systems are comparatively new but in fact many early forms of communication at a distance used this method. The runner with spoken or written message was not digital but any method in the least way technical was. Even coding has existed for many hundreds of years back to the times of the early Greeks and Romans.

Electrical binary methods also are not new, they date back to the beginning of telegraphy in the eighteen hundreds. Our modern 1 and 0 have their earlier parallels in the *dash* and *dot* of the Morse Code (after Samuel F.B.Morse, the American artist and inventor) also in the *mark* (M) and *space* (S) of the telegraph code (in early systems a pen made marks or left spaces on a moving paper tape).

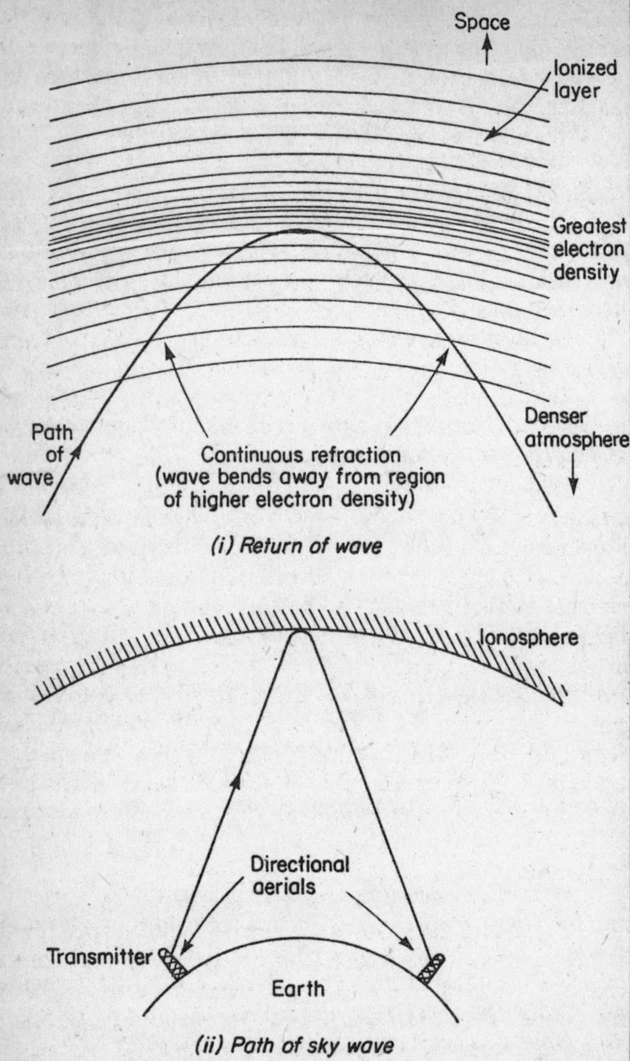

Fig. 4.17 Wave refraction by ionosphere

4.3.1 Telegraphy

Communication resulting in a printed record at the receiving end by use of *teleprinters* is still much in use and because the rate of information flow is restricted to approximately the maximum speed at which an operator can type, bandwidth requirements are low.

The teleprinter is not unlike an electric typewriter except that it is linked by a telegraph transmission channel (2-wire line and/or multiplex system) to a distant teleprinter and use of either keyboard results in the printing of characters on the remote machine with a local copy if required. The first systems which printed on a narrow paper tape have mainly given way to page-printers (as with a typewriter) and *commands* are typed at one end to cause the paper to be moved for commencement of a new line (*carriage-return* and *line-feed*) on the other machine.

When a key is depressed a binary signal representing the letter, numeral or command of the key is transmitted to line. A 5-bit code is used giving $2^5 = 32$ different combinations, (4/A1.2.2) almost doubled by using two special commands *letter shift* and *figure shift*. Following the depression of either of these keys, all letters or all figures are printed until the other is pressed. For example the code 11001 (in telegraph language MMSSM) prints the letter W if a letter shift signal was earlier sent but a figure 2 if the latest shift signal was for figures. Thus an operator wishing to transmit W2 and unaware which shift signal is currently effective would depress the following keys:

Letters, W/2, Figures, W/2.

Fig. 4.18(i) and (ii) show two forms of the line signal for the key W/2 as an example, also the time scale; these are for the character itself. In operation every character is preceded by a *start* pulse of the same length as a single code bit and followed by a *stop* pulse of 1½ times this length. Typically a single bit occupies 20ms, hence one character is transmitted in 150ms.

At the receiving end the printing mechanism is held at rest by

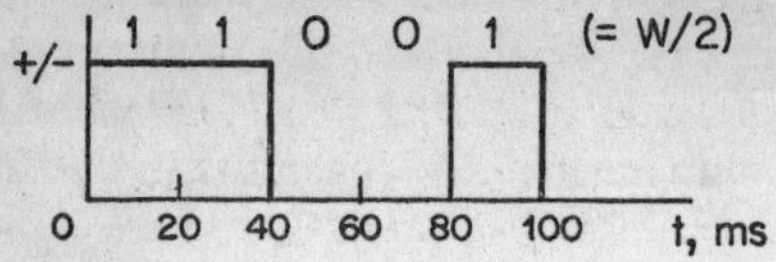

(i) Single-current (unipolar)

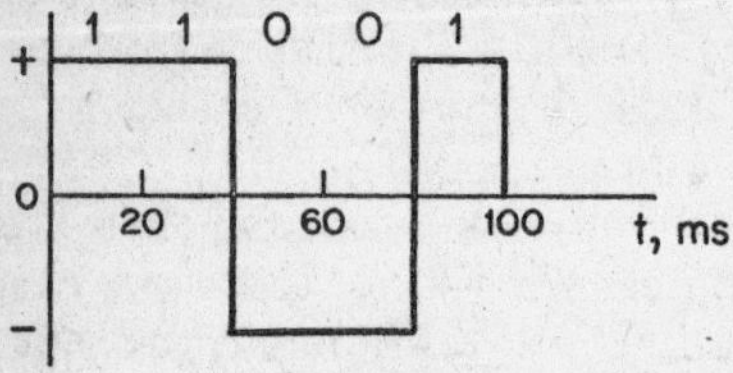

(ii) Double-current (polar)

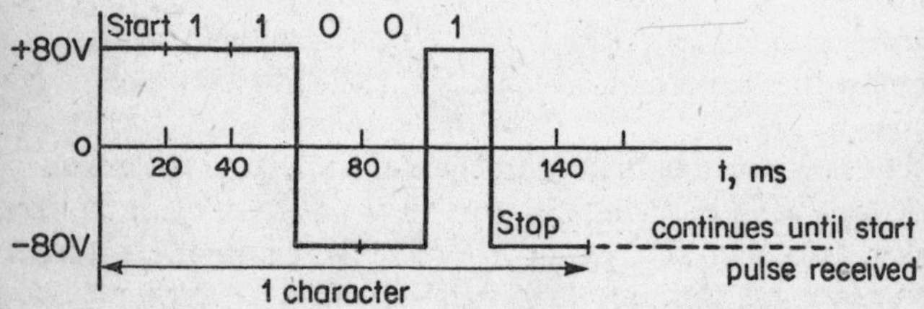

(iii) Double-current with start and stop pulses

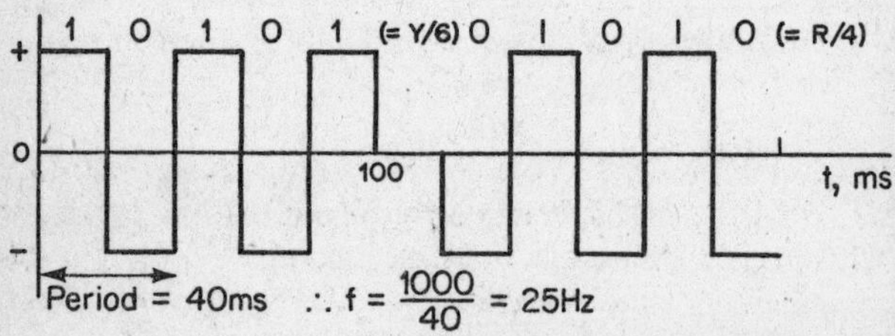

(iv) Characters with maximum number of change-overs

Fig. 4.18 Teleprinter signals

the –80V (other voltages are also used) incoming line signal [see (iii)] until the signal swings positive on the arrival of a start pulse. The 5-unit code then causes the appropriate character to be selected, which one of the two characters for a particular code depending on the previous shift (letters or figures). Using the start-stop method ensures that the receiving mechanism is in synchronism with the transmitting machine at the beginning of each character. When a code such as for "carriage return" (00010) or "line feed" (01000) is received no character is printed but the paper carriage or roller is moved accordingly.

If we work on the basis of one word needing on average 5 letters and one space (6 characters), then since each character requires 150ms, one word occupies 900ms, equivalent to a maximum rate of 66.67 words/minute which is normally ample. To prevent faster typing beating the receiving mechanism, the keyboard is locked while each character is being transmitted, a feature not noticed at typing speeds below the maximum.

At 20ms per bit, the transmission rate is a maximum of 50 bits/sec. This in telegraph parlance is referred to as 50 *bauds* (after the French telegraph engineer, M.E.Baudot). From the ideas on information theory in Section 1.5, another basic formula can be developed, $C = W \log_2 n$, C again representing channel capacity and W, the bandwidth. n is the number of signal states (in this case 2, i.e. 0 and 1).

$$\therefore \text{ for } n = 2, C = W \log_2 2 \quad \text{i.e. } C = W$$

indicating that a 50 baud (bits/sec) circuit can be accommodated within 50 Hz bandwidth. This of course, is for a system with perfect filters and no noise, such is far from the case in practice and it has been found that a bandwidth of at least 120 Hz is normally required.

This can be looked at in another way through Fourier. If we imagine a series of alternate 1's and 0's as occur in the characters R/4 (01010) and Y/6 (10101) [see Fig. 4.18(iv)], this is obviously the greatest rate of change-over, hence as shown in the figure can be represented by a square wave of frequency 25 Hz. A bandwidth of this value would be required for a sine wave but to accommodate the 3rd and 5th harmonics(2/1.4.2) for a sufficiently square wave for telegraphy, a bandwidth of 25 x 5 = 125 Hz is needed. In practice as shown above many multiplex systems work at approximately this value, i.e. 120 Hz per channel. Theoretically therefore

$$\frac{3400 - 300}{120} \approx 25$$

telegraph channels can be multiplexed on a single commercial speech channel, 24 or 18 are standard arrangements. A typical multi-channel voice-frequency (MCVF) telegraph system commences with Channel 1 carrier frequency at 420 Hz, Channel 2 at 540 Hz (120 Hz spacing) up to Channel 24 at 3180 Hz. The elements of such a system are shown in Fig. 4.19 for one end of a teleprinter circuit working over Channel 3. In this case the teleprinter works on an *earth-return* basis, all power supplies having the pole not shown connected to earth. The send channel filter ensures (i) that frequencies outside of the channel band are not transmitted to cause interference in other channels, (ii) that the modulator of one channel has no effect on other channels. The receive channel filter simply selects its own band of frequencies from the 18 or 24 different bands delivered by the receive line.

Modulation is discussed further in Chapter 5. Most likely to be used is amplitude or frequency modulation. For telegraphy the former simply implies transmitting the carrier frequency to line for a 1 (mark), none for a 0 (space). With frequency modulation two different frequencies represent 1 and 0, e.g. on Channel 1 450 Hz and 390 Hz (carrier frequency ± 30 Hz). This is sometimes known as *frequency-shift* keying (f.s.k.) and is less sensitive to impulse noise because the demodulator only operates on specific frequencies.

4.3.2 Regeneration

Although most digital systems other than telegraphy could

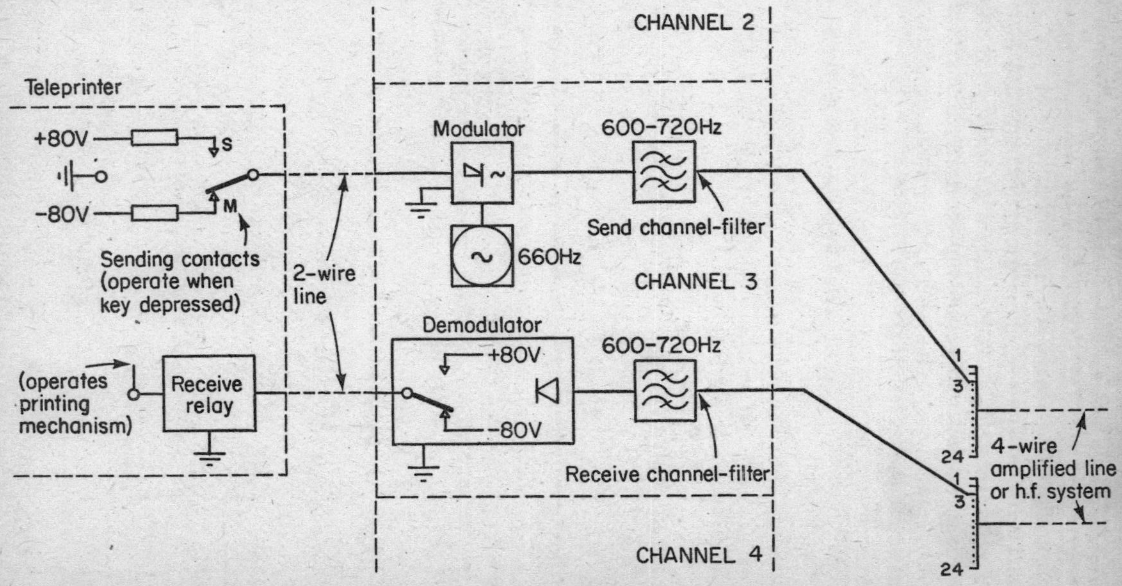

Fig. 4.19 Channel 3 of M.C.V.F. telegraph system.

also be classed as "high frequency" systems and they are carried by both line and radio, here we consider them separately because of the inherently different technique. Multiplex digital systems are the most recently developed and it must be evident from our knowledge of computers that binary transmission has such operational advantages that they are not only very much in fashion but are destined to encroach upon many traditionally analogue fields in the future. Let us look at one of the secrets of success first.

Fig. 4.20 shows a line signal which contains two positive pulses, each lasting for about 0.5μs. The pulses have noise impressed upon them and even without the noise we should find that they are not as square-topped as we would wish because as Fourier analysis shows,(2/1.4) unless the channel bandwidth is infinite the pulse will have sloping sides. This is also a reminder that for pulse transmission we may need a large bandwidth. The channel noise level is high, much higher than could be tolerated by an analogue signal of the same

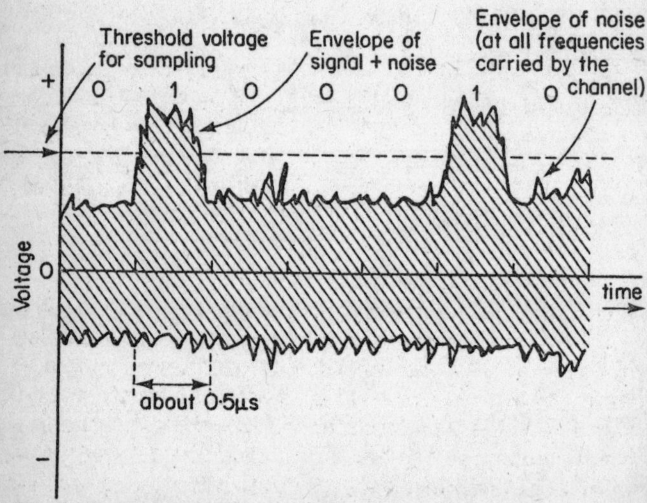

Fig. 4.20 Reception of digital signals with noise present

peak voltage as for the pulses. The systems already examined have amplifiers spaced along the route, digital systems have *regenerators* instead, so called because they detect an incoming pulse and generate a new one, the original being discarded and the "clean" and undistorted new pulse transmitted onwards. The difference between amplification and regeneration is therefore that with the former, noise is amplified and passed forward together with any signal impairment, with regenerators noise is blocked and the on-going signal is as good as the original (there are slight imperfections but these are of no great significance to us yet).

The regenerator may have a sampling threshold voltage as shown in the figure. If at the instant of sampling the signal voltage is above the threshold a pulse (1) is generated at the output, if the voltage detected is below the threshold, no pulse (0) is sent. Irrespective of the length of circuit therefore, the pulses at the output of the receiving-end regenerator should be identical with those transmitted from the sending-end. Failure occurs should there be a "spike" of noise voltage above the threshold at the instant of sampling a 0 when it is then interpreted as a 1.

This enhanced immunity from noise applies to the particular type of pulse modulation known as *pulse-code modulation (p.c.m.)* which is being used on an increasing scale because of its superiority over other pulse systems, we therefore concentrate on this one in particular. The principles are studied in more depth in Chapter 5.

4.3.3 Pulse-Code Modulation Systems

Regenerators, like amplifiers are one-way devices so the system is 4-wire and because most p.c.m. line systems are multiplex with some 24 - 30 bothway speech channels per system, terminating sets are needed to combine both directions of each channel. Whereas we have so far considered f.d.m. systems we are now looking at one based on *time-division multiplexing* (t.d.m.). The difference between the two methods is that f.d.m. allocates a different fraction of the whole bandwidth to each channel continuously but t.d.m. allocates the whole bandwidth to each channel but only for a fraction of the time. Thus the line is available only to Channel 1 for a given period of time, then to Channel 2, followed by Channel 3 etc. When the last channel has been served, the allocation repeats. Exposing all channels to the line once constitutes a *frame* which contains *time slots*, one per channel. It is perhaps easier to appreciate what happens by assigning times to the events so, assuming commercial speech channels and a sampling rate of 8 kHz (as we saw in Chapter 1, at least twice the highest signal frequency and explained in more detail in Chapter 5), we might imagine a sampling method as in Fig. 4.21(i). We are considering a 32-channel arrangement as this is one of the world standard systems, it has 30 speech channels plus 2 for operational purposes. Switch S rotates 8000 times in 1 second, i.e. 125µs per revolution, hence "looking at" each channel for a little less than 125/32 = 3.91µs. During each 3.91µs *sampling time* the analogue signal voltage on the channel input is measured and passed to the *encoder*. For one revolution of switch S (the *sampler*) the encoder receives the information for 1 frame. The encoder changes from analogue to digital[4/4.8] so if we assume Channel 1 to have at the instant of sampling a signal voltage of 35 mV, this applied to the encoder results in an output pulse train 00100011 as shown in the figure [35 = $(0 \times 2^7) + (0 \times 2^6) + (1 \times 2^5) + (0 \times 2^4) + (0 \times 2^3) + (0 \times 2^2) + (1 \times 2^1) + (1 \times 2^0)$[4/A1]]. At this instant Channel 1 is said to be occupying its time slot. If, to give us more practice with binary numbers, we also imagine the Channel 2 signal to have a value of 69 mV (01000101), Channel 3 250 mV (11111010) and Channel 32 41 mV (00101001), the signal output from the encoder during a frame would be as in Fig. 4.21(ii). Frame 2 is shown commencing with Channel 1 at its next sampled value 125µs after the first at, say, 29 mV (00011101). These line signals are regenerated regularly along the route and finally at the receiving end.

Because the attenuation of a pair of wires in a cable increases smoothly with frequency mainly due to the shunting effect of wire-to-wire capacitance, we cannot sensibly talk about bandwidth of a cable as we do a filter which has a sharp frequency

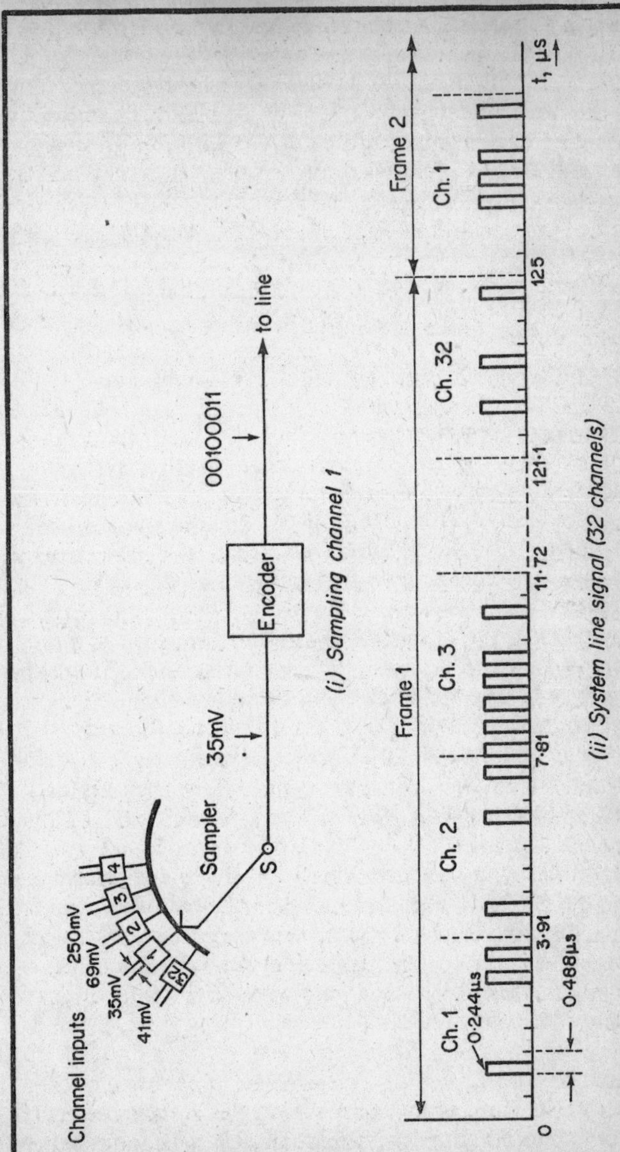

Fig. 4.21 PCM sampling and encoding

cut-off. Other features must therefore be quoted such as the value of the attenuation at which the loss is considered excessive. Also the bandwidth required for satisfactory p.c.m. transmission in a particular case is not easily determined but at least we can calculate the bit rate which for our 32 channel system works out as follows:

Number of bits per frame = 8 x 32 = 256

there are 8000 frames per sec,
therefore line bit rate = 256 x 8000

= 2,048,000 b/s = 2.048 Mb/s.

and as a very rough guide we could use the same figure for bandwidth, say 2 MHz. However most practical systems manage with much less, even down to half this amount. At the highest frequency in the band the attenuation of a pair of wires in a junction or toll cable approaches 20 dB/km, and the length which is generally tolerated before regeneration becomes necessary is no more than about 2 km, quite short compared with the distance between amplifiers in a f.d.m. system. The main compensating factor is that the regenerators are small, easily accommodated in manholes in an underground route or even on poles and require only moderate d.c. power. The power need not be provided locally since it can be fed over the phantoms of the two pairs of wires, the method can be deduced from Fig. 4.8(ii).

At the receiving-end a decoder performs the digital-to-analogue conversion(4/4.8) to regain the original sample values which are then distributed to the appropriate channel demodulators for reconstruction of the original signal waveforms. A schematic of the arrangement is shown in Fig. 4.22. Let us run through what happens again for revision using the figure as a guide:

(i) when a channel is *active*, that is, speech frequencies exist on the terminating set 2-wire terminals, a sample is taken at the send terminals every 125μs (at 8 kHz).

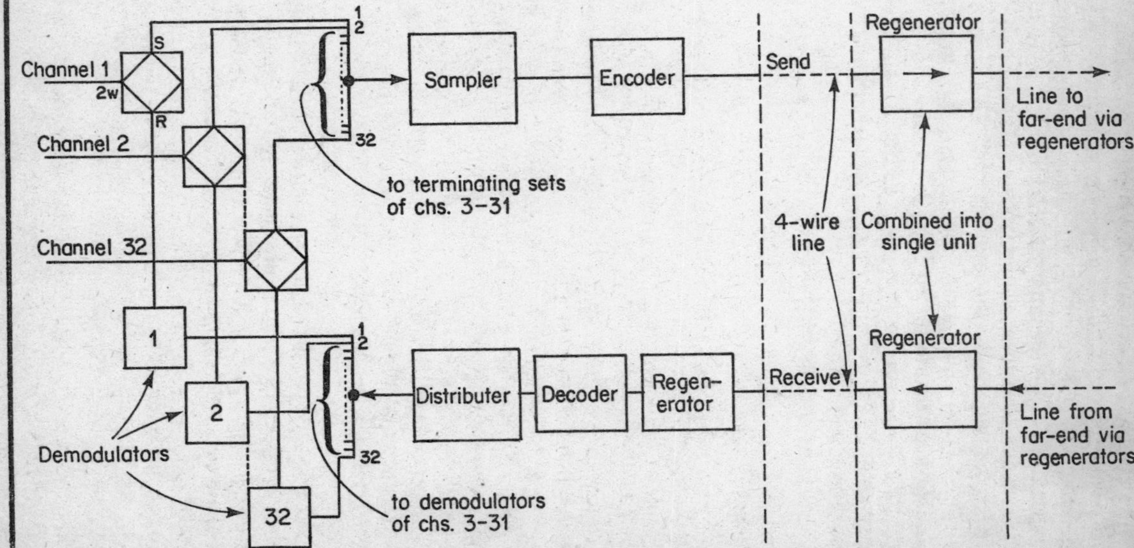

Fig. 4.22 Elements of a 32-channel PCM system.

(ii) the sample, which is in the form of a voltage measurement of the speech waveform at the instant of sampling is passed to the encoder

(iii) the encoder puts out the appropriate binary code in the form of 0.244μs pulses onto the send line. All pulses are of the same amplitude (usually between 2 and 3 volts).

(iv) the sampler switches to the next channel and as above, the binary code corresponding to the voltage present at the send terminals of the terminating set is sent to line, so building up a line pulse pattern as in Fig. 4.21(ii). The pulses do not occupy fully their allotted time slots, in the figure for example they are shown as utilizing half. If the channel time-slot is 3.91μs, then each of the 8 bits occupies

$$\frac{3.91}{8} = 0.488\mu s, \text{ with a pulse width } \frac{0.488}{2} = 0.244\mu s.$$

The effects of inductance (L) and capacitance (C)(2/2) are that current and voltages in these components take time to change from one value to another, thus because L and C occur in pulse circuits (especially in filters) a pulse can neither rise to its full value nor fall back to zero in no time, there is always some delay, to this extent Fig. 4.21(ii) gives a wrong impression. If the trailing edge of a pulse (the fall from maximum to zero) is delayed excessively it encroaches on the rising edge of the following pulse and the time separation between two adjacent channels is impaired. The overlapping of pulses which have suffered distortion is considered in Chapter 5.

(v) on pair-type cables a regenerator is likely to be required after about 2 km, this receives the attenuated pulses and line noise and generates new pulses which are then passed onwards.

(vi) at the far-end, receiving equipment similar to that shown in Fig. 4.22 first decodes each 8-bit train of pulses and then via the distributor passes the analogue voltages which occur every 125μs, into the appropriate channel demodulator

(vii) From the demodulator the speech waveform appears and

is fed to the 2-wire side of the terminating set via its receive terminals.

These are general ideas only. Certainly the sampler would not be a mechanical switch as Fig. 4.21 suggests and one feature of major importance not yet mentioned is that of synchronization between the two ends. Briefly this is accomplished by a controlling oscillator at the sending-end locking the receiving-end into synchronism by signals over a channel reserved for this purpose. Signalling takes up another channel so effectively the system caters for 30 speech circuits.

A cursory glance only but sufficient for appreciation of the essential features. Having some idea of the system itself is of great help when we look more closely at some of the techniques on which it is based. It is difficult perhaps at this stage to appreciate that conversation over a p.c.m. link consists solely of strings of pulses, all of the same amplitude.

4.3.3.1 Higher-Order Systems

As long as each pulse in a digital system can be recognized in a sufficiently short time, it is possible to take several p.c.m. systems together, (say 4), lock them into alignment, detect all corresponding pulses in each system and form them into new t.d.m. groups running at 4 times the bit rate of the component systems. This is illustrated in Fig. 4.23(i). The equipment which combines systems is called a ***multiplexing unit***(m.u.x.)

The first four digits of Channel 1 of each of the four systems are shown with a pulse-width equal, for example, to half the pulse time-slot. The time scale in the figure is not one in use but is chosen for convenience. When assembled as at the bottom of (i) in the figure it is evident that the m.u.x. bit rate is four times that of the individual systems since within their 4μs time slots the m.u.x. transmits four pulses. Thus within the 16μs displayed in the figure the m.u.x. pulses carry the information of all four systems. This leads to a suggested hierarchy as shown in (ii) where each m.u.x. assembles four of the next lower order to combine four times the number of channels at four times the bit rate. In this case the more

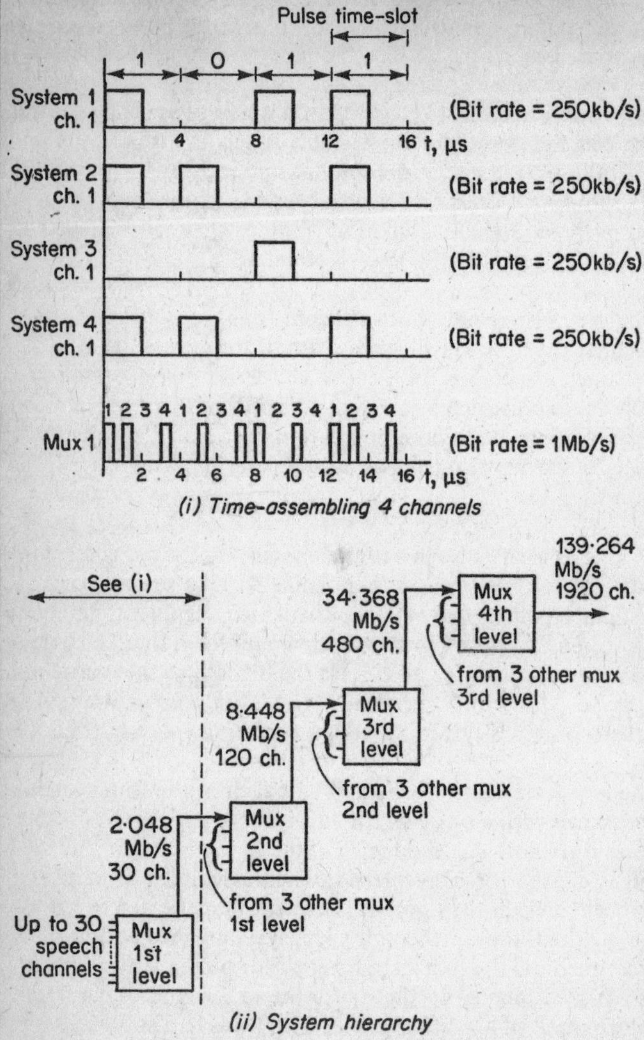

Fig. 4.23 Multiplexing for higher order PCM systems

practical figures for bit rate are shown, they do not progress exactly in multiples of four because allowance is made for "housekeeping" signals which are for controlling the complex and fast-operating circuits at both ends of the link.

Figures shown for the numbers of speech channels do not tell the complete story. In fact any m.u.x. input which can handle the bit rate required may be connected directly to a computer high speed data transfer system, television channel or even a f.d.m. supergroup.

The final bit stream at about 140 Mb/s is carried by a coaxial cable, microwave or satellite link. Even higher bit rates are possible over optical fibres (see Chapter 7).

5. SIGNAL PROCESSING

Processing implies the treatment of a signal for a particular purpose such as making it suitable for transmission over some path or to gain the economy of multiplexing. Usually therefore, unless transmission is at the baseband frequencies a signal is processed immediately before it is launched onto a cable or into the atmosphere, ultimately, when delivered by the transmission path, reverting by a second processing action to its original form. We can recognize therefore that the processing of greatest importance to us in this book is (i) modulation, (ii) frequency translation of a modulated waveform, (iii) demodulation.

Any sine wave can be described by the mathematical expression V sin ωt or V sin $(\omega t + \phi)$ where ϕ represents a constant phase angle. A little revision may be appropriate first. The frequency f is in cycles per second (Hz). In one cycle we consider the wave to be "turning" through 360° or 2π radians.(2/1.2) The number of radians per second or angular velocity, denoted by ω is therefore 2π x f and the total number of radians a wave turns through after a reference starting time (usually at t = 0) is rads./sec x time in secs, i.e. ωt which is usually how the X-axis of the graph is labelled.

Now a sine wave is pictured simply by a graph of sin ωt on the Y-axis against ωt on the X-axis. If it is moving positively through zero value at t = 0 then it is plotted as at (i) in Fig. 5.1 and represented by V_1 sin ωt where V_1 is the maximum value (the sines of angles have only values from 0 to 1). To plot waves of the same frequency but different phases we only need to shift them by the phase-difference as shown in the figure at (ii). The phase-difference (ϕ radians) is added to the equation, giving V_2 sin $(\omega t + \phi)$ and this indicates that the wave is always ϕ radians in advance of a reference one.

Considering a general carrier waveform Vsin $(\omega t + \phi)$, any of the parameters V, ω or ϕ can be varied so as to carry information, giving amplitude, frequency or phase modulation respectively.

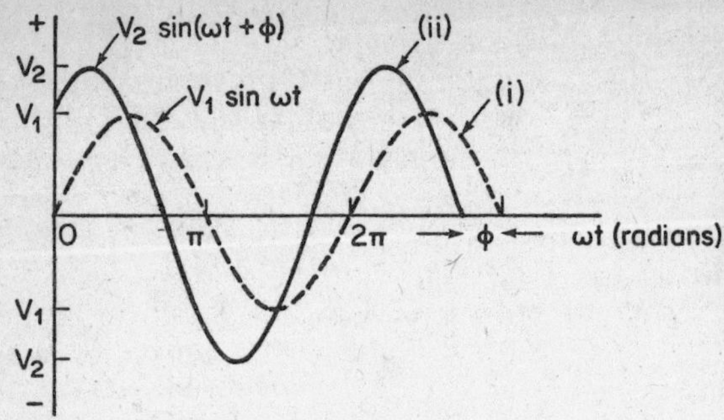

Fig. 5.1 Two sine waves with a phase difference

From Section 4.2.3.1 it is evident that in all modulation processing, the carrier frequency, which we will denote by f_c, is very much greater than the modulating frequency, f_m.

5.1 ADDITION OF TWO SINE-WAVES

Before becoming involved in the extensive subject of modulation we ought first to look at a principle which may be mistaken for this because it also involves two waves of different frequencies. This is when with two waves together, neither modulates the other, they simply add. We only consider two sine waves of equal amplitude but slightly different frequencies (f_1 and f_2 where $f_1 > f_2$) to appreciate the basic process. By so doing we also take our trigonometry one step further.

Consider the two voltages being added as shown at the top of Fig. 5.2. They are of the same amplitude V volts but $f_1 > f_2$. If at regular small time intervals the two curves are added and the result plotted as at the bottom of the figure, by inspection we see that the result is a new frequency

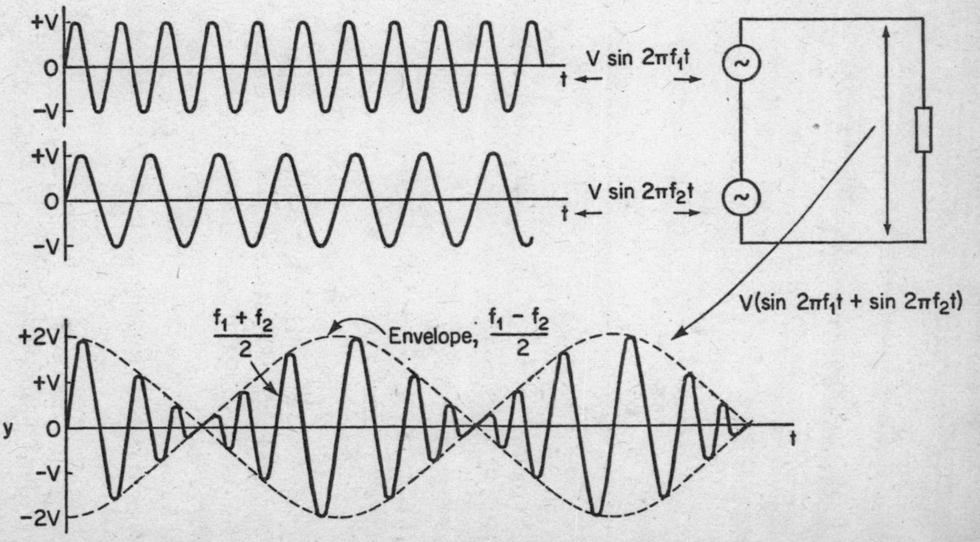

Fig. 5.2 Addition of two sine waves

$\frac{f_1 + f_2}{2}$ but with an amplitude envelope at $\frac{f_1 - f_2}{2}$

The mathematical proof is not too difficult if we extend our knowledge of trigonometry first from Appendix 3.

The equation to the resultant waveform is required, let the voltage of the wave be denoted by y. Then

$$y = V(\sin 2\pi f_1 t + \sin 2\pi f_2 t)$$

and from Appendix 3, Section A3.2, equation (vi)

since $\sin C + \sin D = 2 \sin \frac{C + D}{2} \cos \frac{C - D}{2}$, then

$$y = V\left(2 \sin \frac{2\pi f_1 t + 2\pi f_2 t}{2} \cdot \cos \frac{2\pi f_1 t - 2\pi f_2 t}{2}\right)$$

$$= 2V\left(\sin 2\pi \frac{f_1 + f_2}{2} \cdot t \cdot \cos 2\pi \frac{f_1 - f_2}{2} \cdot t\right)$$

which represents a frequency of

$\frac{f_1 + f_2}{2}$ multiplied by a second one $\frac{f_1 - f_2}{2}$

Cosine in the second term indicates that the wave is sinusoidal in shape but commences 90° later as shown at $t = 0$ on the graph. Also from the graph we see that the amplitude of the new frequency

$\frac{f_1 + f_2}{2}$ is in fact varying between its two extremes at

twice the frequency $\frac{f_1 - f_2}{2}$, i.e. at $(f_1 - f_2)$, this is known known as the *beat frequency*.

Thus with the straightforward addition of two sine waves in a linear circuit (i.e. Ohm's Law always applies, there are no non-linear elements such as diodes present) a new frequency equal to half the sum of the component frequencies is produced, varying in amplitude at a beat frequency equal to the difference between them. Hence two audio-frequency oscillators may be synchronized by mixing their outputs and detecting the beat frequency on headphones or an a.c. meter, one is then adjusted in frequency until the beat is zero whereupon the two oscillator frequencies are equal.

This also explains in greater detail the radio carrier interference problem mentioned in Section 4.2.3.4 for conditions where the beat note is audible.

If the two waves have such a frequency difference that, for example, one is a multiple of the other, then the wave of lower frequency simply suffers distortion (second, third harmonic etc.)(2/1.4.1) This occurs because the waves are in phase once per cycle of the lower frequency.

5.2 AMPLITUDE MODULATION

At this stage, having looked at transmission systems first, we are in a position to study the technicalities of modulation with some good appreciation of its purpose. This section looks at amplitude modulation, used from the earliest days of radio and still much in use today.

We begin as usual with sine waves. The carrier, f_c is most likely to be a sine wave in any case, but the modulating wave, f_m, is not. To carry information it is likely to be a band of frequencies and much more complex. Nevertheless to consider a single frequency for f_m is practical (time signals for example are given in "pips' of tone) and it enables us to see what happens without getting too involved in mathematics. In the preamble to this chapter, f_c is stated as being very much greater than f_m, say, 30 up to several hundred times, in our drawings however, purely so that what happens is pictorially clear, we use a ratio of only 5 e.g. f_c = 100kHz, f_m 20 kHz as in Fig. 5.3. This shows the modulating wave f_m at (i) and from t = 0 to t = 50μs the unmodulated carrier wave at (ii),

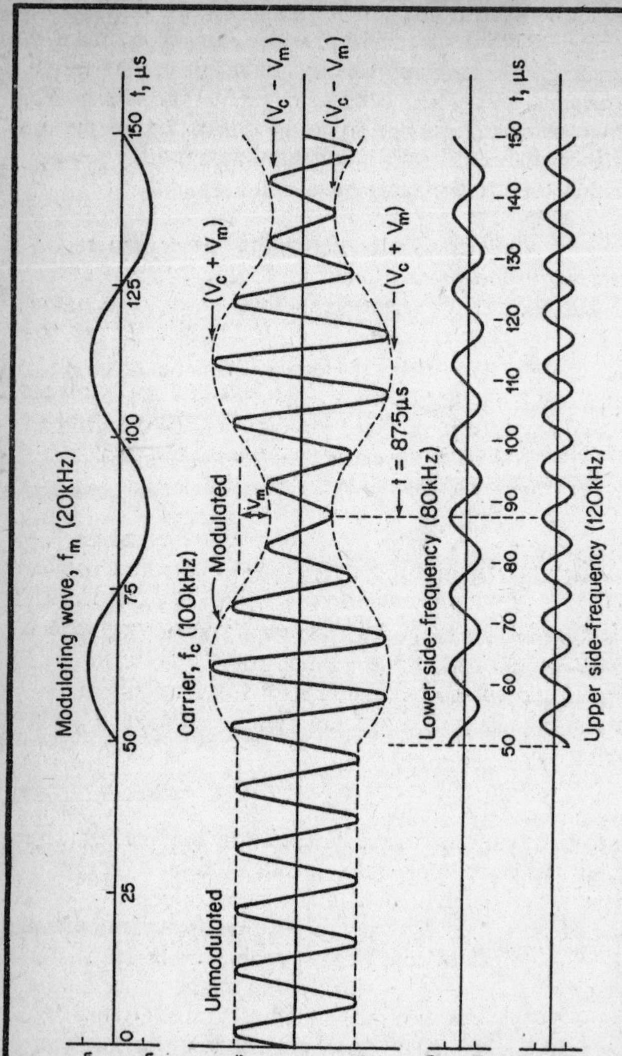

Fig. 5.3 Amplitude modulation

simply a sine wave at 100 kHz. From $t = 50\mu s$, modulation takes place and f_m is impressed on the amplitude of the carrier. If f_m has an amplitude V_m, then the carrier amplitude, originally V_c, varies between $(V_c + V_m)$ and $(V_c - V_m)$ as shown. The carrier now contains the information from the modulating wave. Clearly V_m should not exceed V_c because the latter cannot fall below zero.

Analysis of the modulated carrier shows that it in fact comprises three different frequencies added together, the carrier itself, f_c, a *lower side-frequency* $(f_c - f_m)$ and an *upper side-frequency* $(f_c + f_m)$. the latter are shown at (iii) and (iv).

Let the instantaneous value of the unmodulated carrier wave = v_c, then $v_c = V_c \sin \omega_c t$ volts where $\omega_c = 2\pi f_c$ rads/s and for the modulating wave $v_m = V_m \sin \omega_m t$ volts where $\omega_m = 2\pi f_m$ rads/s.

If the amplitude of the carrier wave is varied by the modulating wave as shown in Fig. 5.3(ii), then the instantaneous value of the modulated wave, v, varies in two ways (i) by the effect of v_m and (ii) by the fact that the wave itself is also varying sinusoidally, i.e.

$$v = \underbrace{(V_c + V_m \sin \omega_m t)}_{\text{(i)}}\ \underbrace{\sin \omega_c t}_{\text{(ii)}} \quad \text{volts}$$

$$\therefore\ v = V_c \sin \omega_c t + V_m \sin \omega_m t \cdot \sin \omega_c t \quad \text{volts}$$

The last part of the expression can be expanded by recourse to equation (v) in A3.2, giving

$$v = V_c \sin \omega_c t + \frac{V_m}{2}\left\{\cos(\omega_c - \omega_m)t - \cos(\omega_c + \omega_m)t\right\} \text{volts}$$

i.e. $v =$

$$V_c \sin \omega_c t + \frac{V_m}{2} \cos 2\pi(f_c - f_m)t - \frac{V_m}{2} \cos 2\pi(f_c + f_m)t \text{ volts,}$$

the three waves already mentioned. The first, $V_c \sin \omega_c t$ shows that the original carrier is present. No part of the expression contains ω_m hence none of the original modulating frequency remains.

$$\frac{V_m}{2} \cos 2\pi (f_c - f_m)t$$

is shown as the lower side-frequency $(f_c - f_m)$ in Fig. 5.3(iii) and because it is a cosine wave it leads the carrier at $t = 50\mu s$ by $\pi/2$ radians i.e. $90°$ (a wave does not start at maximum as suggested in the figure of course, it is shown like this to emphasize the phase relationship). Similarly

$$-\frac{V_m}{2}\cos 2\pi (f_c + f_m)t$$

is the upper side-frequency $(f_c + f_m)$ shown at (iv), it is a negative cosine wave and hence lags by $90°$.

If at any instant therefore the amplitude of the unmodulated carrier is added to those of the side-frequencies, the result, taking polarity into account, gives the amplitude of the modulated carrier at that instant. Let us make one simple check on this at, say, $t = 87.5\mu s$ where it would appear from Fig. 5.3(ii) that the modulated wave amplitude is $-(V_c - V_m)$.

The carrier itself from $t = 50\mu s$ has moved through 3 complete cycles plus $3/2\pi$rads $(270°)$. Its amplitude is therefore $V_c \sin 270° = V_c \times -1 = -V_c$. The lower side-frequency has moved through exactly 3 cycles, hence with a value $V_m/2$. The upper side-frequency has moved through 3½ cycles again having a value $V_m/2$ (remember that for the side-frequencies the amplitudes are relative to their values at $t = 50\mu s$). Thus by adding,

$$-V_c + \frac{V_m}{2} + \frac{V_m}{2} = -V_c + V_m, \text{ ie } -(V_c - V_m).$$

The difference between adding two sine waves (Section 5.1) and modulating one with the other now becomes a little more clear. With addition, the two separate frequencies can subsequently be regained by filtering, with modulation which is the *product* of two waves, this is not possible because f_m has disappeared in the modulation process.

When modulation is by a band of frequencies rather than a single one, then each individual frequency within the band receives the treatment analysed above. If the band extends from f_1 the lowest frequency to f_2 the highest, the modulated carrier will no longer look neat and tidy as in Fig. 5.3(ii) but have a complex amplitude variation (remember, in practice there are many cycles of f_c to even the highest modulating frequency f_2). Our analysis for the single frequency is sufficient however for an understanding of the process, the main difference being that the two new components in the modulated wave are *lower and upper sidebands* (l.s.b. and u.s.b.), the single frequency having given way to a band. The modulated wave therefore consists of (i) the carrier, (ii) the lower sideband extending from $(f_c - f_2)$ to $(f_c - f_1)$, (iii) the upper sideband extending from $(f_c + f_1)$ to $(f_c + f_2)$. These features are better displayed on a frequency spectrum diagram as we see next.

5.2.1 The A.M. Frequency Spectrum

The three component frequencies are displayed on an amplitude/frequency basis in Fig. 5.4(i) for a single modulating frequency f_m. Typical frequency figures are added.

At (ii) is shown the spectrum when modulation is by a band consisting at the particular instant of 8 separate frequencies ranging from f_1 (lowest) to f_2 (highest). The l.s.b. is said to be *inverted* because the side-frequency $(f_c - f_1)$ is higher than $(f_c - f_2)$ whereas in fact f_2 is higher than f_1. Conversely the u.s.b. is said to be *erect*. Typical frequency figures added show how a radio transmission on 900 kHz appears when modulated by a band of audio frequencies 100 – 4500 Hz. The total transmission bandwidth is $(f_c + f_2) - (f_c - f_2) = 2f_2$, that is, twice the highest modulating frequency so that although modulation is from a 4.4 kHz band, the modulated carrier occupies 9 kHz. This is known as *double sideband* working (d.s.b.) and is normally used for radio broadcasting with each radio transmission taking up 9 kHz of the available frequency spectrum. On medium waves, for example, the number of kHz

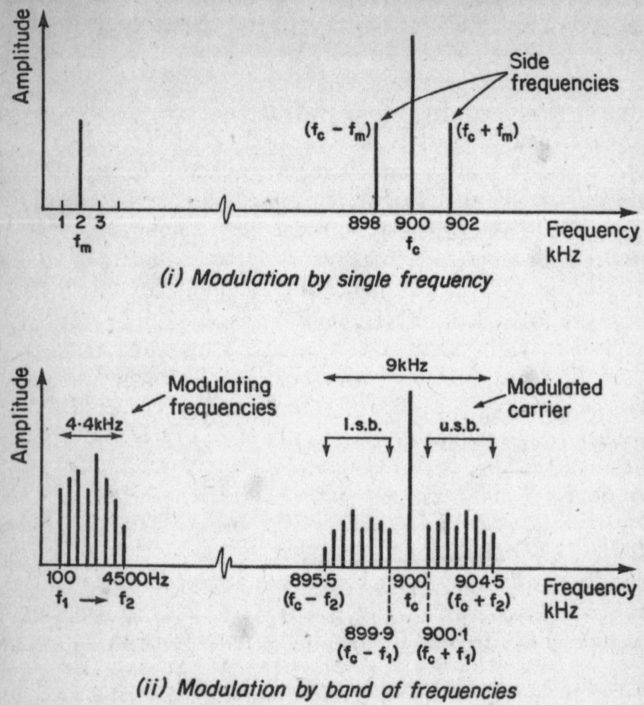

Fig. 5.4 Spectra of amplitude modulated waves

available is limited hence the somewhat crowded condition of the band. In addition radio receivers ideally must not only tune to f_c but be able to accept 4.5 kHz on either side, no more.

We consider in Section 5.2.3 the obvious question as to why both sidebands are needed for radio broadcasting whereas a single one is the backbone of most multiplex systems.

5.2.2 Modulation Factor

This is a means of expressing the degree of depth to which a carrier wave is modulated. In Fig. 5.3(ii) the amount by which the modulating wave penetrates the carrier is V_m and the factor (m) by which it does this relates to the value of the carrier itself, hence

$$m = \frac{V_m}{V_c}$$

When m = 0 there is no modulation, when m = 1 there is full modulation for then $V_m = V_c$ and V_c falls to zero, that is, the carrier wave dips right down to the time axis while alternately rising to $2V_c$. In the figure

$$V_m = \frac{V_c}{2}, \text{hence } m = \frac{V_c/2}{V_c} = 0.5$$

The modulation factor may also be expressed as a percentage hence

$$\text{percentage modulation} = m \times 100 = \frac{V_m}{V_c} \times 100\%$$

5.2.3 Modulation Techniques

Each frequency in a modulating signal has its counterpart in both of the sidebands generated in the modulation process, it follows therefore that the total information contained in the modulating signal resides in each sideband. When bandwidth is at a premium it is therefore desirable to transmit one sideband only and we have already encountered widespread use of this principle in Chapter 4. The carrier may also be suppressed because it contains no information and also requires heavier duty amplifiers to handle it. But there is one major difficulty when eventually demodulation is required, the carrier has to be reintroduced and we shall see that its frequency should be within a few Hz of the original. This normally requires provision of an oscillator at the receive-end controlled by synchronizing signals from the sending end. In multiplex line and radio systems this is no great burden for a master oscillator to which all channel oscillators are linked can be so controlled, the cost of this, no matter how great, is many times outweighed by the savings in transmission equipment. The obvious advantage of s.s.b. compared with d.s.b. in its lower requirement of bandwidth makes the former the clear choice for multiplex systems.

To provide a carrier frequency oscillator in every radio receiver and control it from the transmitting station is not so practical and for this reason ordinary broadcasting includes the carrier with the result that demodulation in the receiver remains a comparatively simple process.

5.2.3.1 Modulators

In Book 3(3/A4.2) we had some practice in discovering the mathematical equations to certain curves, particularly those relating to transistor input characteristics and diodes. These all show how current varies with voltage for a particular device and if the graph is not a straight line the characteristic is said to be *non-linear*. Certain of these belong to a family known as *square-law*, so named because at least one component of the current is proportional to the *square* of the voltage. The general curve equation is in the form of a series but as usually found, only the first few terms are needed, i.e.;

$$i = a + b\nu + c\nu^2 \ \ldots$$

and a typical practical transistor input characteristic is shown in Fig. 5.5 where this is the relationship which applies from $\nu = 0.7$ to $\nu = 0.75$ volts and the values of the constants give

$$i = 8.745 - 25.26\nu + 18.267\nu^2 \ \ldots \text{amps}$$

Earlier we decided that modulation arises from the *product* of two waves and by developing the square law formula further we can see what happens. In modulators f_c and f_m are applied in series, hence

$$\nu_c = V_c \sin \omega_c t, \ \nu_m = V_m \sin \omega_m t \ \text{and} \ \nu = \nu_c + \nu_m .$$

$$\text{Then} \ i = a + b\nu + c\nu^2 = a + b(\nu_c + \nu_m) + c(\nu_c + \nu_m)^2$$

$$\therefore \ i = a + b(\nu_c + \nu_m) + c(\nu_c^2 + 2\nu_c\nu_m + \nu_m^2)$$

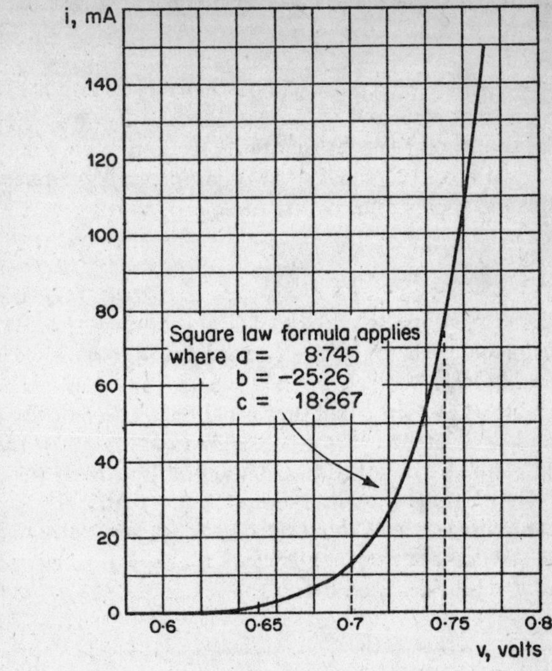

Fig. 5.5 Typical transistor input characteristic

$$\therefore\ i = a + bV_c \sin \omega_c t + bV_m \sin \omega_m t + cV_c^2 \sin^2 \omega_c t$$

$$+ 2cV_cV_m \sin \omega_c t \cdot \sin \omega_m t + cV_m^2 \sin^2 \omega_m t.$$

Now, developing the 5th term by use of equation (v) of A3.2, this becomes

$$c\, V_cV_m \left\{ \cos(\omega_c - \omega_m)t - \cos(\omega_c + \omega_m)t \right\}$$

and by simply changing the order of the terms:

$$\underset{d.c}{i = a} + \underset{carrier}{bV_c \sin\omega_c t} + \underset{mod.freq.}{bV_m \sin\omega_m t}$$

$$\underline{\textit{l.s.b.}} \qquad\qquad\qquad\qquad \underline{\textit{u.s.b.}}$$

$$+ cV_cV_m \cos 2\pi (f_c - f_m)t - cV_cV_m \cos 2\pi (f_c + f_m)t$$

$$\underline{\textit{(carrier)}^2} \qquad\qquad \underline{\textit{(mod.freq.)}^2}$$

$$+ cV_c^2 \sin^2 \omega_c t + cV_m^2 \sin^2 \omega_m t$$

The modulation components (carrier, l.s.b and u.s.b.) are there, but so are others. The last two components we have difficulty in recognizing but by use of equation (vii) of A3.2 they become

$$c\cdot\frac{V_c^2}{2}(1 - \cos 2\omega_c t) \text{ and } c\cdot\frac{V_m^2}{2}(1 - \cos 2\omega_m t)$$

and on removing the brackets each is seen to contain a d.c. component and one effectively at double the original frequency. These and the other unwanted ones are removed by using a band-pass filter for the modulation products only, the d.c. components, modulation frequency and its square (at $2f_m$) are well below the filter cut-off, the (carrier)^2 component (at $2f_c$) is well above. On the frequency spectrum of Fig. 5.6 the components are displayed to show how correct filtering is an uncomplicated process.

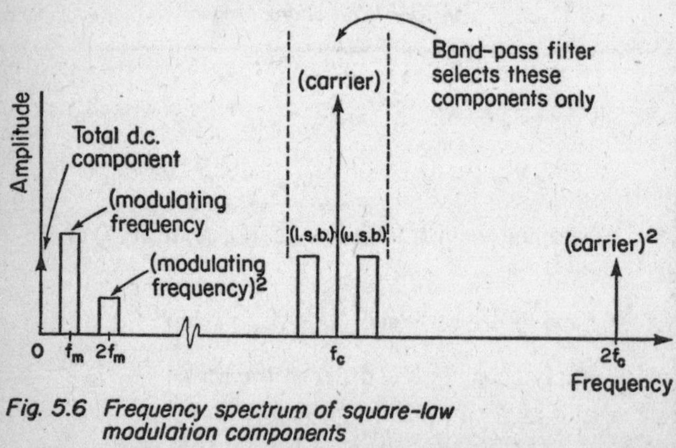

Fig. 5.6 Frequency spectrum of square-law modulation components

Summing up this section so far it is now clear that to build a modulator we need only to apply both carrier and modulating frequencies in series to a transistor working on a square-law input characteristic. Fig. 5.7 shows the elements of such a circuit, using an npn transistor. The carrier is applied directly to the base and the modulating frequency to the emitter circuit, both are therefore in series in the base-emitter (input) circuit. R_1, R_2 and R_3 together fix the d.c. bias(3/3.2.4) so that the transistor operates over the square-law part of its characteristic. All frequency components (Fig. 5.6) are generated in the collector circuit with the $L_1 C_1 L_2 C_2$ tuned transformer acting as a band-pass filter to select carrier, l.s.b. and u.s.b. only.

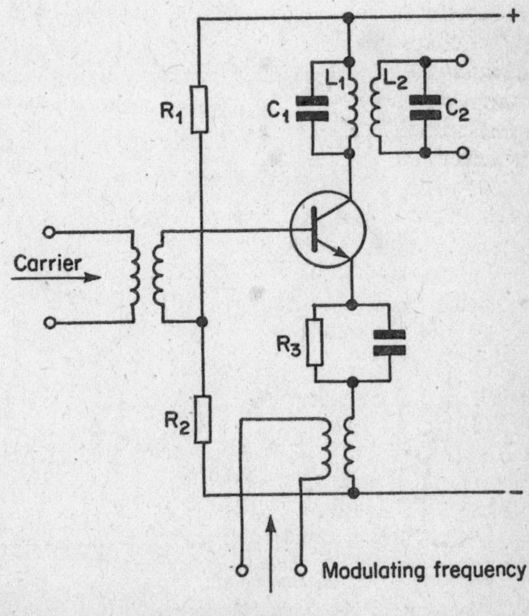

Fig. 5.7 Single stage modulation

Carrier suppression is not easily obtained by filtering for it only differs from the sidebands by the lowest modulating frequency, below 100 Hz or a few hundred at the most, hence a sharp cut-off is necessary. The problem is more easily solved by balancing out the carrier in the modulator first. Fig. 5.8(i) shows Fig. 5.7 in schematic form, by doubling up as in (ii) a balanced modulator is derived and whatever the direction of the carrier currents at any instant (e.g. as shown by the arrows), as far as the transformer feeding the load R_L is concerned, they cancel out, thus with the circuit halves balanced no carrier voltage appears across R_L. On the other

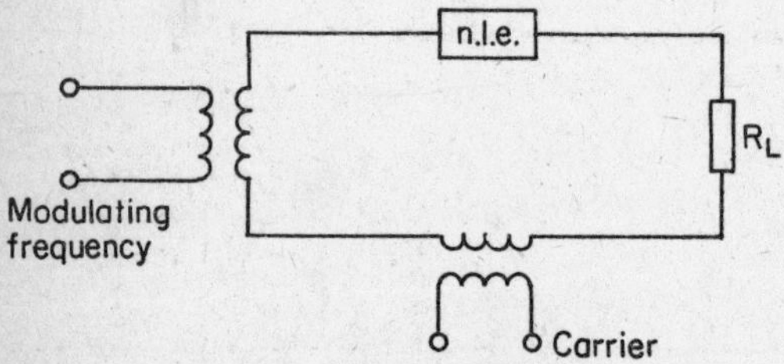

(i) Unbalanced

n.l.e. = non-linear element, e.g. diode or suitably biassed transistor.

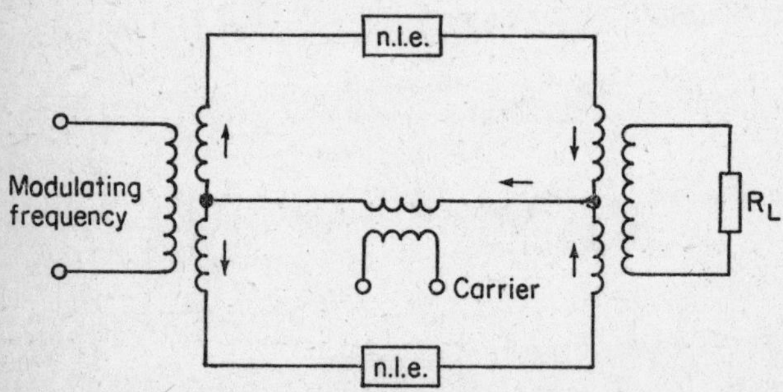

(i) Balanced

Fig. 5.8 Unbalanced and balanced amplitude modulators

hand the modulating frequency is transmitted to R_L provided that the non-linear elements (usually diodes or transistors as in Fig. 5.7) have low impedance. The carrier voltage is sufficient to alternately switch the diodes "on" and "off", hence "chopping" the modulating frequency at the carrier frequency as shown in Fig. 5.9(iv). At (i) is the carrier and at (ii) what might be called the switching waveform produced by its action

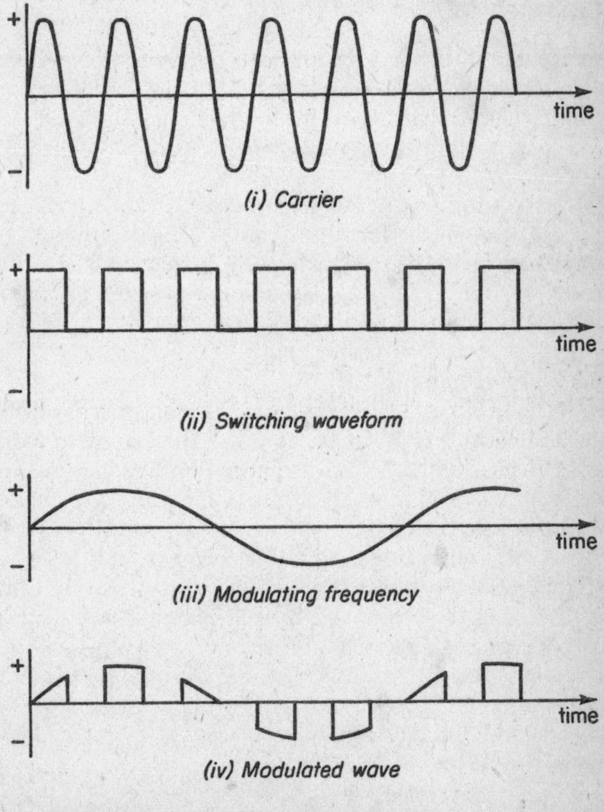

Fig. 5.9 Waveforms in a balanced modulator

on the diodes which are switched on and off once per carrier cycle. The modulating frequency at (iii) therefore appears as bursts of voltage on each alternate half-cycle of the carrier, (iv) is in fact the product of (ii) and (iii). We need not go into the mathematics because the basic principle is similar to that above. Through Fourier analysis (2/1.4.2) a square waveform as in (ii) may be resolved into its component sine waves, the series for the waveform shown being:

$$\frac{1}{2} + \frac{2}{\pi}\left(\sin \omega t + \frac{\sin 3\omega t}{3} + \frac{\sin 5\omega t}{5} + \ldots\right)$$

The product of this series with the equation to the modulating frequency again produces a $\sin \omega_c t.\ \sin \omega_m t$ component, the output in fact consisting of the modulating frequency, l.s.b., u.s.b. and harmonics, but no carrier.

This is the simple balanced arrangement, more complicated ones are used including the *double-balanced* or *ring* modulators in which the modulating frequency is also suppressed. Double sideband operation is converted to single-sideband by means of a band-pass filter.

5.2.4 Demodulation

Recovery of the original modulating frequency is, as might be expected, known as *demodulation*. It is the inverse of modulation. The term *detection* is frequently found as meaning the same but in general this only refers to the process as applied to radio waves, thus an h.f. system uses demodulators whereas radio receivers have detectors. There are of course many different techniques and circuits, two examples only follow but through these we are able to appreciate the general principles. We look at detection first because, as already mentioned, broadcast receivers accept carrier and both sidebands and detection in this case is a relatively straightforward process with uncomplicated circuitry.

5.2.4.1 Diode Detection

Referring back to Fig. 5.3(ii), the modulating frequency is seen to be impressed on both the positive swings of V_c and the negative. Only one of these is needed for recovery of the information hence all positive or all negative half-cycles are first dispensed with. For this the semiconductor diode has many advantages, not the least being its low capacitance which avoids passage of r.f. current when it is in the reverse state. The effect of a diode and its resistance load R on the modulated wave of Fig. 5.3(ii) is therefore as shown in Fig. 5.10(i), (ignore capacitor C at present).

Although the waveform is now fully positive or negative-going it clearly consists of pulses of current at carrier frequency, these must be removed but the shape of the envelope retained. This is effected by use of a capacitor in a similar fashion to the use of a reservoir capacitor in power rectification,(3/3.1) in fact the techniques are very similar only the order of frequency and therefore component values are different. The diode is therefore followed by capacitor C as shown in Fig. 5.10(ii) with the output at terminals 1 and 2 (which is also the voltage v_c across C) as in (iii). In (iii) the pulses of current at the output of the diode as in (i) are shown dotted, these are applied to C so from t = 0 C commences to charge,(2/2.2) in a rather complex manner because not only is it supplied by a varying voltage but it is also shunted by R which continually drains current away. During the first pulse of current (lasting in our example for 5μs), C does not reach the full voltage of the pulse before the latter falls. During the gap from 5-10μs, C is discharging through R. As soon as the voltage of the second pulse exceeds v_c, current recommences to flow into C and increases its charge (10 – 15μs). Between 15 and 20 μs C is again discharging through R. This process of adding charge on each pulse and its leaking away between pulses reaches equilibrium and continues until after t = 50μs. The pulses then rise in value and following the same principles v_c also rises (remember that in practice there are many more cycles of the carrier per cycle of the modulating frequency than can be shown in the drawing). When the pulses fall in value, v_c follows them down because of the drainage of current by R (e.g. 75 – 100μs). The values of C and R must be carefully chosen so that v_c follows the modulating frequency waveform as closely as possible.

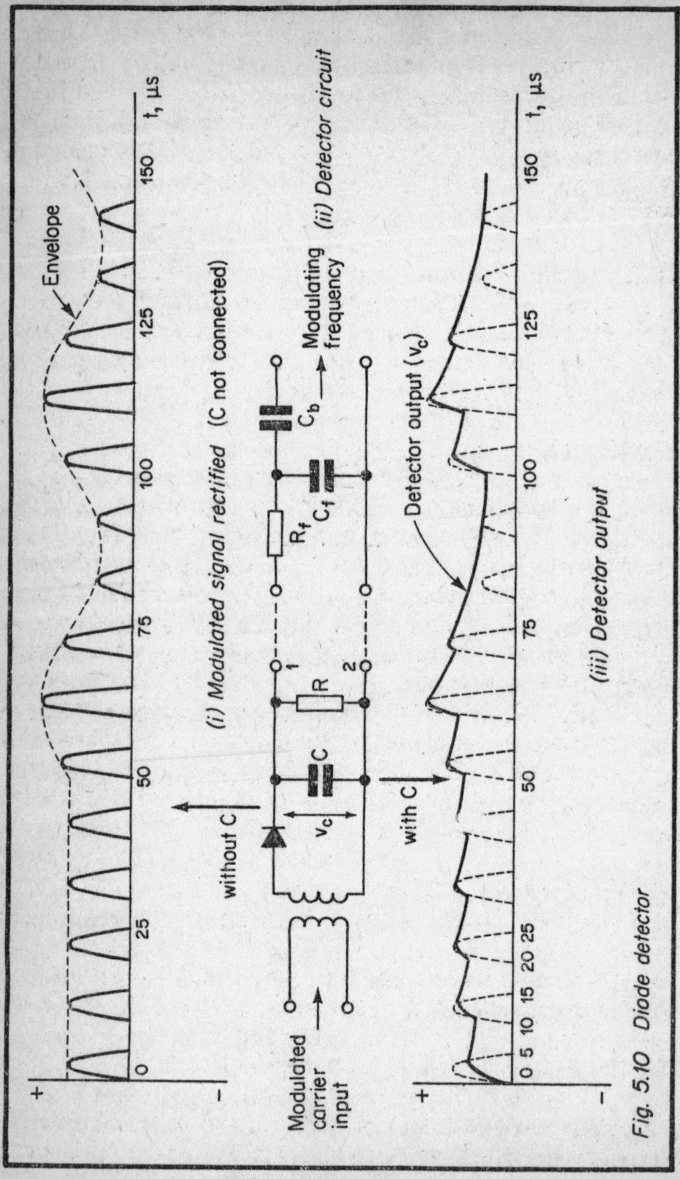

Fig. 5.10 Diode detector

Comparing the detector output curve at terminals 1 and 2 with the original modulating frequency of Fig. 5.3(i), we seem to be a long way from perfection. Firstly it is all on one side of the time axis because the wave has been rectified and has a d.c. component. A series capacitor is the simplest method of correcting this by blocking the d.c. yet passing the modulating frequency, it must therefore have low reactance to the latter.

With this the wave then swings equally above and below the time axis. Secondly the lack of smoothness in the graph of v_C is at the carrier frequency, this may be filtered out by a second capacitance-resistance combination. Both the above are shown in (ii) of Fig. 5.10 as C_b for the blocking capacitor, and C_f, R_f for the filter components.

5.2.4.2 The Balanced Demodulator

Because most h.f. systems embody the single sideband suppressed carrier (s.s.b.s.c.) technique, we ought to examine this particular one, especially as it can be done without getting too involved with mathematics. It has already been indicated that suppressed-carrier systems require reinsertion of the carrier in the demodulation process, it will become clear as we progress that without doing so the sideband cannot be brought back down to baseband. Consider the use of a balanced modulator as in Fig. 5.8(ii). For clarity let f_m represent the whole band of modulating frequencies and if therefore the single sideband, for example, the lower $(f_c - f_m)$ is applied to the modulating frequency terminals then the modulator output into R_L becomes

mod.freq. *l.s.b.* *u.s.b.*

$$(f_c - f_m) + \{f_c - (f_c - f_m)\} + \{f_c + (f_c - f_m)\} + \text{harmonics}$$

The l.s.b. reduces to f_m and the u.s.b. to $(2f_c - f_m)$ thus the whole expression can be rewritten as

$$(f_c - f_m) + f_m + (2f_c - f_m) + \text{harmonics.}$$

A simple low-pass filter can select f_m and reject all other components which are much higher in frequency. Adding this filter and the carrier supply to the balanced modulator produces a *balanced demodulator*.

The above assumes that f_c for the incoming sideband is an identical frequency to the f_c supplied locally to the demodulator. Suppose however that they are not the same, say f_{c_1} and f_{c_2}, then the (lower) single sideband is given by $(f_{c_1} - f_m)$ and the desired demodulation product by

$$f_{c_2} - (f_{c_1} - f_m) = f_{c_2} - f_{c_1} + f_m,$$

that is, f_m is now increased or decreased by $f_{c_2} - f_{c_1}$, the difference between the two carriers. As mentioned above, f_m represents a band of frequencies hence all frequencies in the band are shifted by this amount. For speech 10 – 15 hz is found to be the maximum shift tolerable, for data transmission however, only ± 2Hz is allowed. Such close alignment would be a rigorous requirement for two separate oscillators thus frequently the technique is used of transmitting a low level of unmodulated carrier from the sending-end of a multiplex system to lock the receiving-end demodulator oscillator to the correct frequency. The system is then not a truly suppressed carrier but virtually so.

5.2.5 Frequency Conversion

We briefly met frequency conversion (i.e. frequency translation or mixing) in Section 4.2.3.2 when discussing microwave systems. The very high frequency-modulated carrier is brought down to a lower frequency mainly because amplifiers for the higher frequencies are more complicated and expensive. As an example, to amplify a modulated wave at 4 GHz, it is first converted down to a carrier wave at 70 MHz but carrying the same sidebands. Amplification follows at 70 MHz, then a second stage of frequency conversion back up to 4 GHz. This is not modulation although similar circuits and principles apply, it is simply a process of shifting a modulated wave in frequency without change to the modulation.

A commonly used example of frequency conversion is in broadcast radio receivers. A receiver tunes in any one of a range of carrier waves, so to avoid the difficulty of providing amplifiers capable of operating over the whole range, any frequency to which the receiver is tuned is first converted to an *intermediate frequency* (i.f.) by a *frequency-changer* stage. Amplification then follows at i.f. and the amplifiers only need to operate over a band wide enough to accommodate the sidebands and centred on the i.f. (usually around 470 kHz for the popular long and medium wave receiver). In this case conversion is said to be effected by a *heterodyne* process (from Greek, = different force) and the receiver is classed as a *superheterodyne* type (originally "supersonic heterodyne").

Considering a carrier f_c with its sidebands $f_c \pm f_m$ (again let f_m represent the *band* of modulating frequencies), from the work we have already done it is clear that mixing these with another frequency in a non-linear device (or modulator) produces sum and difference frequencies for each one. Ignoring all those products which are not required and which are eventually filtered out and calling the new (heterodyne) frequency f_h, Table 5.1 shows the products generated. Rearranging the output terms:

$f_{c_1}, (f_{c_1} - f_m), (f_{c_1} + f_m)$
is a modulated wave centred on f_{c_1}.

$f_{c_2}, (f_{c_2} - f_m), (f_{c_2} + f_m)$
is a modulated wave centred on f_{c_2}

and in both cases the original modulation is unchanged, that is, there is merely a frequency conversion from f_c to f_{c_1} and f_{c_2} by mixing with f_h. Of the two modulated waves the one required is selected by a band-pass filter.

Some reality is added by considering a medium wave radio transmission on 909 kHz (f_c) modulated by speech or music having a range of frequencies 100 Hz – 5 kHz (f_m). A radio receiver is therefore tuned to 909 kHz but must equally accept 904 – 914 kHz. Now the frequency-changer stage which converts to the intermediate frequency (470 kHz) contains a variable (heterodyne) oscillator. When this is set to 1379 kHz (f_h) we get, from Table 5.1

TABLE 5.1 PRODUCTS OF FREQUENCY CONVERSION

Input	Heterodyne Oscillator	Output
f_c	f_h	$(f_h + f_c)$ – call this f_{c_1} $(f_h - f_c)$ – call this f_{c_2}
$(f_c - f_m)$	f_h	$f_h + (f_c - f_m) = (f_{c_1} - f_m)$ $f_h - (f_c - f_m) = (f_{c_2} + fm)$
$(f_c + f_m)$	f_h	$f_h + (f_c + f_m) = (f_{c_1} + f_m)$ $f_h - (f_c + f_m) = (f_{c_2} - f_m)$

$$f_{c_1} = 2288 \text{ kHz}, \qquad f_{c_2} = 470 \text{ kHz},$$

and already it is clear that f_{c_1} and its sidebands are to be rejected by filtering so leaving f_{c_2} plus

the lower sideband $(f_{c_2} - f_m) = 465 \rightarrow 469.9$ kHz

the upper sideband $(f_{c_2} + f_m) = 470.1 \rightarrow 475$ kHz

Thus the incoming radio frequency 904 → 915 kHz centred on 909 kHz is converted (or translated) to 465 → 475 kHz centred on 470 kHz.

Fig. 5.11 shows the elements of a typical radio receiver frequency-changer. $L_1 L_2$ is an r.f. transformer coupling the aerial circuit to the base of an npn transistor (pnp is equally usable). R_1, $R_2 C_2$ in conjunction with $R_3 C_3$ provide the correct bias and stabilization. Oscillation is set up in the $L_5 C_5$ tuned circuit(3/3.3) (frequency controlled by C_5) and coupled into the transistor emitter and collector circuits by L_3 and L_4. The coupling between L_3 and L_4 provides the necessary feed-back for oscillation. Thus the incoming radio signal and the local oscillation are "mixed", the desired mixing product being available in the collector circuit and tuned by the *i.f. trans-*

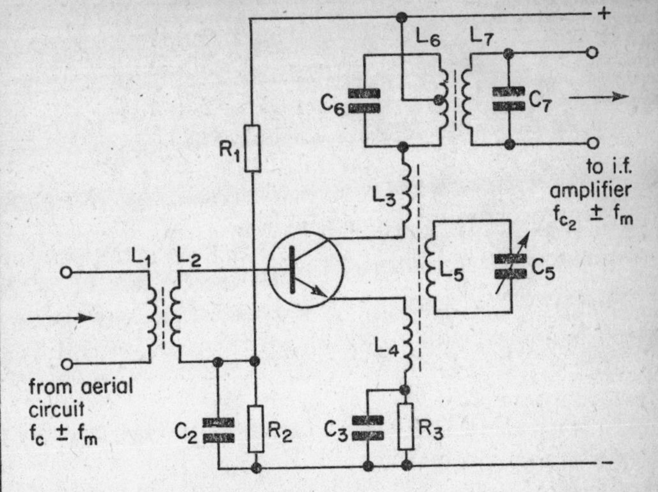

Fig. 5.11 Frequency changer

former L_6C_6 coupled with L_7C_7. One or more i.f. amplifying stages follow, then detection and finally a.f. amplification.

5.3 FREQUENCY MODULATION

Important for its wide use in high quality radio broadcasting, for television sound, microwave and satellite transmission, frequency modulation (f.m.) is typified by constant amplitude but varying frequency. The degree of frequency variation is proportional to the amplitude of the modulating wave whereas the rate of variation is according to the modulating frequency itself.

An important advantage of f.m. arises from the fact that many interference voltages produce amplitude modulated waves, f.m., being insensitive to amplitude changes is unaffected by these whereas a.m. of course, is. F.M., however, requires greater bandwidth, a further example of trading bandwidth for signal/noise ratio, in this case using greater bandwidth to obtain a system more tolerant of noise. The main differences between a.m. and f.m. are apparent on comparison of Fig.5.12 for f.m. with Fig. 5.3 for a.m. The constant amplitude of the f.m. wave is immediately evident, also its variation in frequency, here from +75 kHz on the nominal frequency of 200 kHz to –75 kHz. ±75 kHz is the *frequency deviation* brought about by the variation from 0 to maximum above and below the time axis of the modulating wave. For example, from $t = 20$ to $45\mu s$ the modulating wave is rising to maximum amplitude and the f.m. wave varies accordingly from 200 to 275 kHz, the deviation being proportional to modulating wave amplitude, i.e. to $V_m \sin \omega_m t$ where $\omega_m = 2\pi f_m$.

Now the maximum deviation shown in Fig. 5.12 is 75 kHz, it can in fact be any value we choose, let us give it a general symbol Δf, with its maximum value $\Delta f_{(max)}$ [Δ is the Greek capital letter, delta, frequently used to indicate change, although if this is small, the lower case delta (δ) is used]. Thus since Δf varies according to the amplitude of the modulating wave (e.g. when $V_m \sin \omega_m t = 0$, $\Delta f = 0$, when $V_m \sin \omega_m t = 1$, Δf is maximum) then the frequency deviation is $\Delta f_{(max)} \sin \omega_m t$ and at any instant the frequency of the modulated wave is given by

(nominal frequency + deviation) i.e.

$$f = f_c + \Delta f_{(max)} \sin \omega_m t \quad \text{Hz}$$

where f_c is the nominal (carrier) frequency).

As an example, at $t = 45\mu s$ in Fig. 5.12 the modulating wave has advanced through $25\mu s$ from its commencement at $t = 20\mu s$ $\Delta f_{(max)} = 75$ kHz

$$\therefore \; f = 200 \times 10^3 + 75 \times 10^3 \sin (2\pi \times 10^4 \times 25 \times 10^{-6}) \text{ Hz}$$

$$= 200 + 75 \sin \frac{\pi}{2} \text{ kHz} = 275 \text{ kHz}$$

a simplified calculation but one easily checked in the Figure.

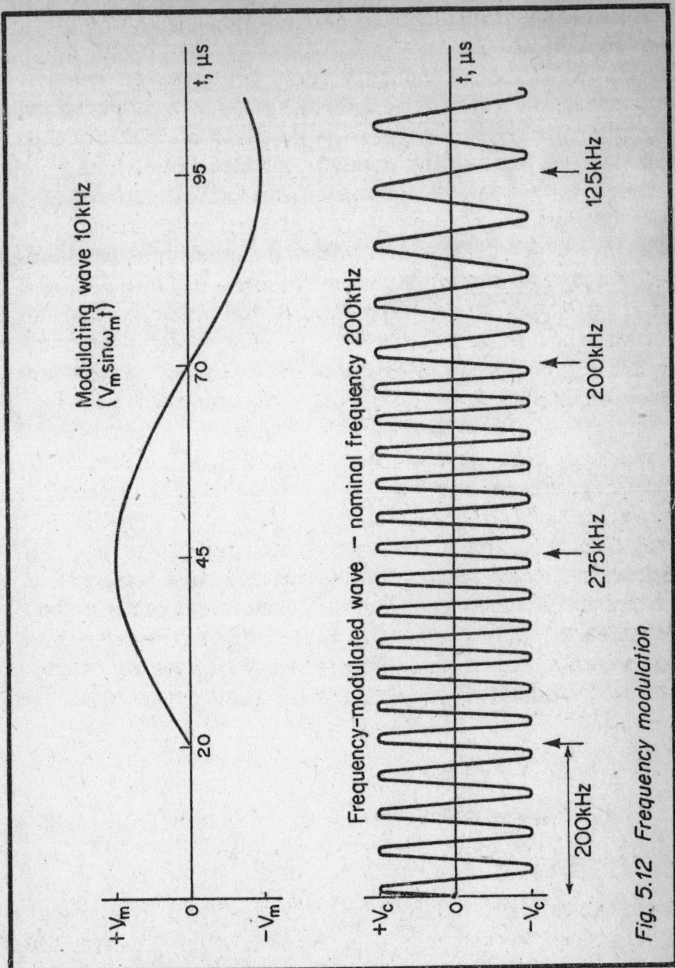

Fig. 5.12 Frequency modulation

For other values of t the instantaneous frequency is similarly calculated.

For our example we have chosen a particular value for $\Delta f_{(max)}$ (75 kHz), a deviation which only occurs at $+V_m$ or $-V_m$. The assumption is that V_m is the greatest amplitude catered for by the particular system with $\Delta f_{(max)}$ built into the design. In other words, for the maximum amplitude of the modulating frequencies a certain *maximum frequency deviation* (sometimes known as the *rated system deviation*) must first be chosen. Theoretically $\Delta f_{(max)}$ can be very large but channel bandwidth limitations are usually overriding.

5.3.1 Bandwidth

Looking again at Fig. 5.12, one might easily conclude that the bandwidth required for the f.m. wave is from 125 to 275 kHz, i.e. $2\Delta f_{(max)}$. This ignores one important factor, however, which is that when the frequency is changing, each cycle differs from those on either side of it and in fact is not truly sinusoidal within itself, being either stretched out or cramped. We now have enough experience to guess that harmonics are generated, hence a greater bandwidth than $2\Delta f_{(max)}$ must be needed.

It would be unwise to seek out the mathematical explanation in this particular case for the going would not be easy unless an appendix of abnormal length to cover some calculus and certain special functions were added. We can, however, take the result of such an exercise which gives the general equation to the modulated wave as

$$v = V_c \sin \left(\omega_c t + \frac{\Delta f}{f_m} \sin \omega_m t\right),$$

v being the instantaneous value of the wave and V_c the maximum value of the carrier. The constant maximum amplitude of the wave is shown by V_c but the variation is a sine function of a quantity which itself also contains a sine term. $\Delta f/f_m$ is the term most usually described as the *modulation index* (M) although other ratios are sometimes used. This immediately confronts us with another complex variation for Δf depends on the *amplitude* of the modulating wave, thus its relationship with f_m continually varies, moreover f_m itself can have any one of a range of values. It is profitless therefore for us to get too deeply involved in such a complex process, hence sufficient to accept without query the general formula:

Bandwidth for the f.m. wave = $2(\Delta f + f_m)$

which is the instantaneous value, i.e. when the modulating wave frequency is f_m and its amplitude is such as to produce a frequency deviation of Δf. Theoretically a slightly greater bandwidth is required but in this formula some higher harmonics of low significance are excluded. The bandwidth constantly changes, resulting in an approximate system bandwidth requirement of $2(\Delta f_{(max)} + f_{m(max)})$ which occurs with the highest modulating frequency at maximum amplitude. Taking the practical example of high quality f.m. radio broadcasting,

$$f_{m\,(max)} = 15\text{ kHz},\ \Delta f_{(max)} = 75\text{ kHz}$$

$$\therefore\ \text{Bandwidth} = 2(75 + 15) = 180\text{ kHz}$$

A *deviation ratio* for a system may also be quoted, it is the ratio between the maximum frequency deviation and the maximum modulating frequency, i.e.

$$\frac{\Delta f_{(max)}}{f_{m\,(max)}}$$

which in the case above is equal to

$$\frac{75\text{ kHz}}{15\text{ kHz}} = 5$$

5.3.2 Modulation Techniques

Evidently to generate an f.m. wave we have somehow to change the frequency of the carrier wave according to the amplitude of the modulating wave. We look at one example of how this can be done and one which also gives us a little revision on tuned circuits. We first recall that a tuned (resonant) circuit operates to the formula

$$f = \frac{1}{2\pi\sqrt{LC}} \qquad (2/3.7)$$

this being the series formula which is also usually sufficiently accurate for parallel circuits. L and C are the inductance and capacitance in henrys and farads respectively. Semiconductor junctions have capacitance,(3/2.4) often a disadvantage but at times useful and for this particular purpose special diodes are manufactured for their voltage-dependent capacitance characteristics when operated with reverse bias. They are usually silicon and are known as variable-capacitance or *varactor* diodes. With reverse bias the diode resistance is extremely high, so much so that the diode virtually presents capacitance only provided that the voltage applied does not swing it into forward bias.

The elements of a modulator circuit are given in Fig. 5.13. L_1 with C_1 and C_2 in series and in parallel with the capacitance of the diode, form a tuned circuit which with the correct reverse bias from V_B and no modulating frequency input, oscillates at f_c. C_3 blocks bias current from the tuned circuit and L_2 is a radio-frequency *choke* (presents a high impedance to f_c but not to f_m) to prevent interaction between the resonant and modulating frequency circuits. When the modulating frequency is applied, the bias on D swings about its *quiescent* value (i.e. motionless, when there is no input signal to disturb it) according to the amplitude of the signal applied. Hence the frequency generated by the oscillator varies accordingly. Because f.m. needs a high bandwidth, transmission is at a comparatively high frequency and changes in diode capacitance of a few pF are sufficient.

This is obviously not the only frequency modulation technique available, it is merely one through which we can most easily see how it can be done. The basic principle of capacitance variation within a tuned circuit however, is probably the most frequently encountered.

5.3.3 Demodulation

An f.m. demodulator translates frequency deviations back into the original amplitude variations of the modulating signal. Since the frequency deviation in the modulated wave is linear with modulating frequency amplitude so too must the demodulation process be linear. Because interference and noise cause amplitude variations in the f.m. signal, it is important

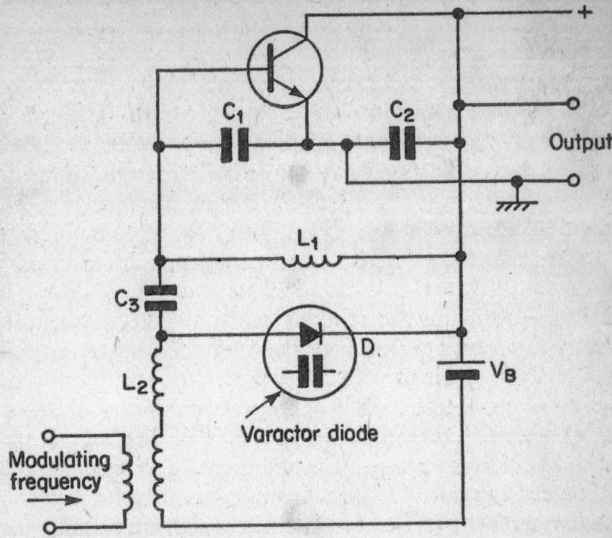

Fig. 5.13 Variable-capacitance diode F.M. modulator

that an f.m. demodulator is affected by frequency only and not amplitude to prevent the noise signal being present in the demodulator output. Certain demodulating circuits do not possess this feature, hence they must be preceded by a *limiter*.

5.3.3.1 Amplitude Limiting

Interference is most easily removed from an f.m. wave by reducing the latter so that all cycles have the same amplitude, by so doing none of the information transmitted as frequency variations is lost, but interference is. We are not examining limiting circuits in detail because the demodulator described in the next section automatically provides its own but some form of limiting must be incorporated otherwise one of the major advantages of f.m. is lost. Limiting may be accomplished for example by use of a transistor operated at a low collector voltage and high load resistance so that the load line falls between cut-off and saturation points(3/3.4.2.1) of sufficiently low values that the wave is *clipped* to a constant amplitude. Many variations of this technique, especially using diodes, are also available.

5.3.3.2 F.M. Detection

Traditionally the names of two American engineers are linked with f.m. demodulation. They are Foster and Seeley and their well-known circuit is of a *phase-difference discriminator*. The original circuit does not provide limiting so has to be preceded by a limiter. Subsequently a *ratio-detector* circuit was developed using similar basic principles but not requiring a separate limiter. However, more recently circuits better suited to IC production because they contain less inductance and capacitance(3/4.2.2) are naturally taking over and the one described below, although very much simplified shows this major basic principle on which many radio f.m. detectors work. All the circuit except for one single LC combination can be absorbed into an IC. Also some IC's, as might be expected, consist almost entirely of components which are suitable for integration but the circuitry becomes quite complex. We therefore look at a circuit which we can understand and from which we can gain a little more design experience. Consider Fig. 5.14(i) in which $L_2 C_2$ forms a tuned circuit resonant at f_C. The bias on the transistor (not shown) and supply voltage are, as suggested in the foregoing section, such that the transistor operates over a load line between saturation and cut-off to provide limiting. From simple use of operator - j(2/1.3.4) and assuming quite reasonably that the resistance of L_1 is very much smaller than its reactance and therefore can be omitted from consideration, we get:

$$i_1 = \frac{v_1}{j\omega L_1}$$

and the induced emf in L_2:

$$e_2 = -j\omega M i_1$$

where M is the mutual inductance between L_1 and L_2

$$\therefore\ e_2' = \frac{-j\omega M v_1}{j\omega L_1} = \frac{-M v_1}{L_1}$$

When the secondary circuit (L_2C_2) is resonant (i.e. f_c is applied), it is purely resistive, with an impedance of, say, $R_2\,\Omega$

$$\therefore\ i_2 = \frac{e_2}{R_2} = \frac{-M}{L_1R_2} \cdot v_1$$

so i_2 is $180°$ out of phase with v_1. The voltage v_2 across C_2

$$= i_2 \cdot \frac{-j}{\omega C_2} = \frac{jM}{L_1R_2\omega C_2} \cdot v_1 = jv_1\left(\frac{M}{L_1R_2\omega C_2}\right)$$

which is therefore $90°$ in advance of v_1 (or with the connexions to L_2 changed over, $90°$ lagging) and we can sum all this up by saying simply that the emitter voltage on the transistor (v_2) is $90°$ (in quadrature) leading or lagging on the base voltage (v_1) according to the circuit connexions.

If the input frequency falls, the L_2C_2 circuit becomes inductive and has a positive angle, if the frequency rises the circuit becomes capacitive and exhibits a negative angle. These changes therefore modify the phase angle existing between base and emitter to values above and below $90°$, hence the period of time in which the base is positive to the emitter (for collector current to flow in an npn transistor) on each cycle. In other words, the frequency deviation of the f.m. signal has been first expressed as a phase-shift then as an amplitude variation. We can see for ourselves just how this happens by constructing the appropriate graphs to show how a phase change between v_1 and v_2 (see Fig. 5.14.(i)) results in an increased collector current time per cycle. The exercise is worthwhile for many types of discriminator incorporate this principle.

With no frequency deviation v_1 and v_2 are $90°$ out of phase (say, for this exercise v_2 is $+90°$ on v_1). The base-emitter voltage depends on $v_1 - v_2$ and we plot this, as in Fig. 5.14(ii) for v_1 at various angles and v_2 $90°$ ahead. As an example, when v_1 has moved through $60°$, its amplitude is proportional to $\sin 60° = 0.866$, v_2 is at $(90 + 60)°$, the sine of which is 0.5, therefore $\nu_1 - \nu_2$ is represented on the graph by 0.866 – 0.5= 0.366. In the figure this has been repeated for frequency deviations giving rise to phase differences of –10° (ν_2 = +80°) and +20° (ν_2 = +110°) . The curves clearly show the amplitude differences and we can make this more practical by assuming some purely arbitrary values for the operating point of the transistor and for saturation and cut-off. Such values are indicated by arrows and dotted lines.We might also choose some frequency simply to bring angles to time, for example, 2.778 MHz for which the periodic time is 360ns. The lower graphs show the collector currents which apply and that they flow for average times as follows:

ν_2 at +90° (no frequency deviation) 92ns
ν_2 at +80° (–10° phase difference) . . . . 78 ns
ν_2 at +110° (+20° phase difference) 106ns

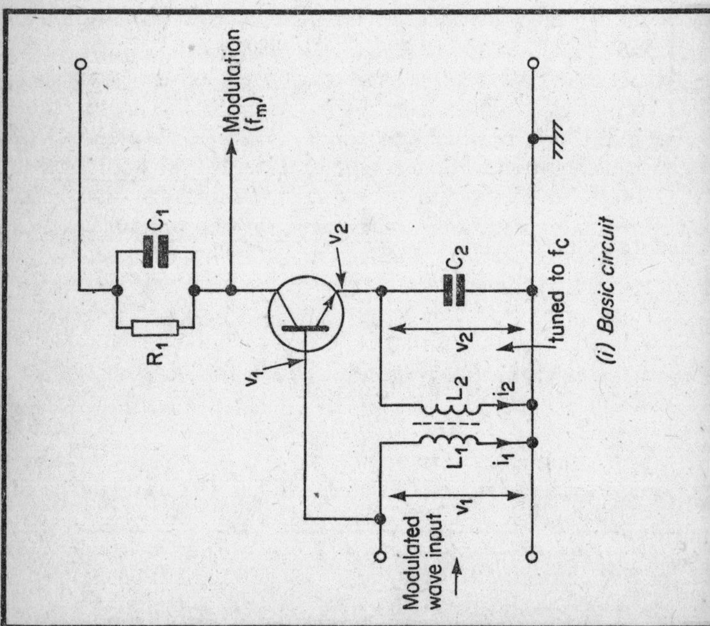

(i) Basic circuit

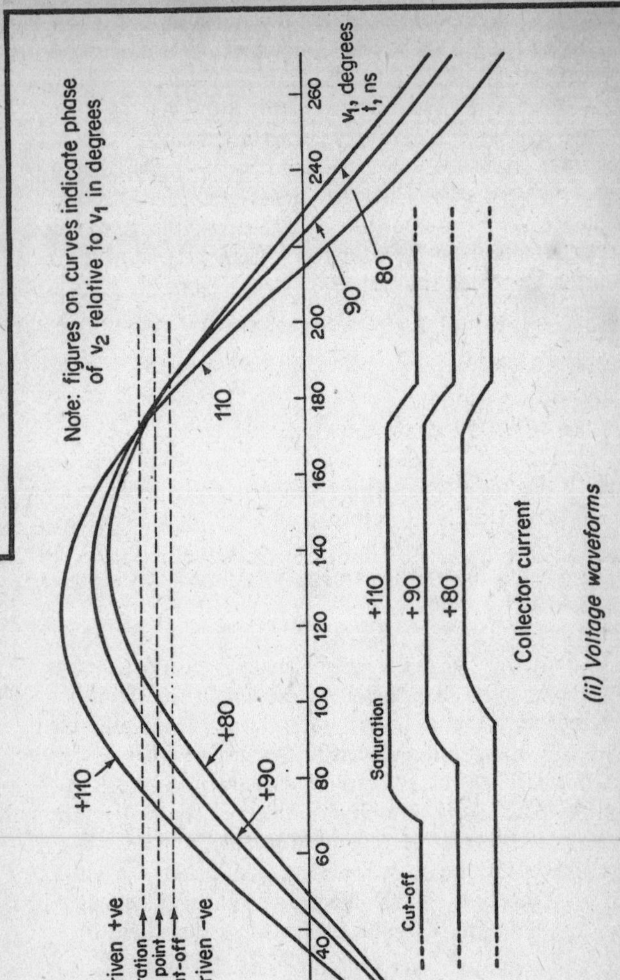

(ii) Voltage waveforms

Fig. 5.14 F.M. quadrature detector

These are the times during which capacitor C_1 in the collector circuit is receiving a charge and recalling the appropriate formula(2/2.2)

$$v_C = \frac{i \times t}{C}$$

it is evident that for greater times the output voltage of the circuit is greater, overall therefore, frequency deviation reappears as amplitude variation.

5.4 PHASE MODULATION

We will be brief with phase modulation (p.m.) because, compared with f.m. it is less widely used. The two types also have many technical similarities, fundamentally because when a wave varies in frequency it is also undergoing phase change. Thus the wave equations have the same form and in fact both are often classed under the same general heading of *angle modulation*.

For p.m. it is the phase deviation which is proportional to modulation amplitude and the rate at which the deviation varies agrees with the modulating frequency. Now quite clearly when we talk about phase-deviation it must be with reference to something. With f.m. the reference is always present at the receiving end, it is the unmodulated carrier, f_c, with p.m. a reference phase has to be generated in the demodulation process and this is one complication which causes f.m. to be used in preference. It can also be shown that it is a slightly less efficient system with regard to utilization of bandwidth.

5.5 PULSE MODULATION

In general in the various modulation processes described above the carrier is a sine wave, one parameter of which is modified according to the amplitude and frequency of the modulating wave. We next study pulse modulation in which the elementary unit when seen on a graph has a rectangular form, known as a pulse. Whereas the sine wave is of one frequency only, the pulse is just the opposite. As we so often remind ourselves,(2/1.4.2) it needs an infinite bandwidth for transmission without distortion, a facility which is just not available. Pulse transmission must therefore always result in some deformation in the original shape but despite this, as seen in the previous chapter, pulse modulation systems have so much to offer that their future expansion is assured. Analysis of pulse waveforms can get very complicated indeed and invariably difficult mathematics are called for. For ourselves we can gain all we need from the construction of just a few graphs.

The graph of voltage or current against time for a train of pulses such as are found in many pulse systems is shown in Fig. 5.15(i). The period T secs. from the commencement of

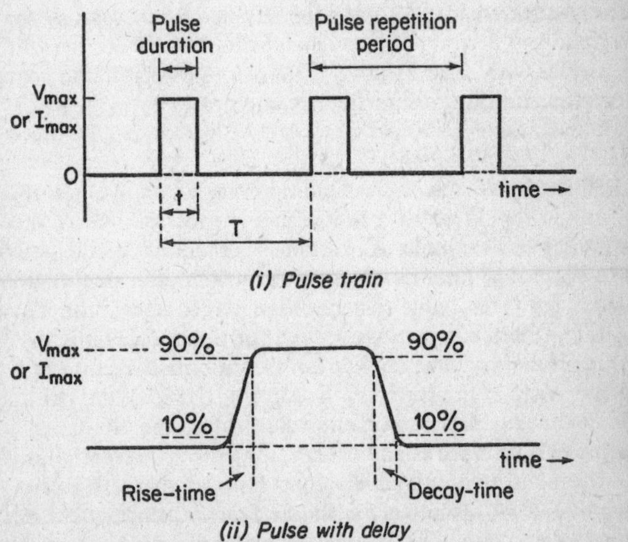

Fig. 5.15 Pulse terminology

one pulse to that of the next is known as the *pulse repetition period*. The *pulse repetition frequency* (p.r.f.) is then f = 1/T Hz. Each pulse lasts for a certain time, known as the *pulse duration* shown in the diagram as t.

For much work, because the pulse neither rises nor falls in zero time as shown in the figure but suffers a delay in so doing, these times need to be quoted. The *rise-time* is that taken for the pulse to rise from 10% to 90% of its maximum value. The *decay time* equally occurs between 90% and 10%. These are shown in (ii) in the figure from which it can be seen that the choice of these two percentages tends to avoid the parts of the curve where most non-linearity exists, these, if included, would lessen the value of comparison in different cases.

5.5.1 Pulse Distortion

There are many purposes for which a pulse is deliberately changed in shape from the rectangular to something completely different[4/4.2.5] but generally in transmission systems change is undesirable although inevitable. We refresh our memories with regard to two main causes of distortion apart from the addition of interference and noise first.

5.5.1.1 RC Networks

Resistance and capacitance abound everywhere in electronics. So to a lesser extent does inductance, hence we look to the former as our example. A resistance-capacitance (RC) network as in Fig. 5.16(i) has a *time-constant* which indicates the time needed for C to charge to a specified percentage of the voltage V_{in}.[2/2.2] Similarly time is needed for the charge on C to decay. Hence a pulse applied to such a circuit results in a voltage V_{out} of the form typified in Fig. 5.16(ii), the rise and decay times depending on the values of R and C. As an example, with both R and C high, the current, severely limited by R, takes a comparatively long time to charge the large capacitance. Alternatively with low R and C a heavy current flows to charge quickly a small capacitance.

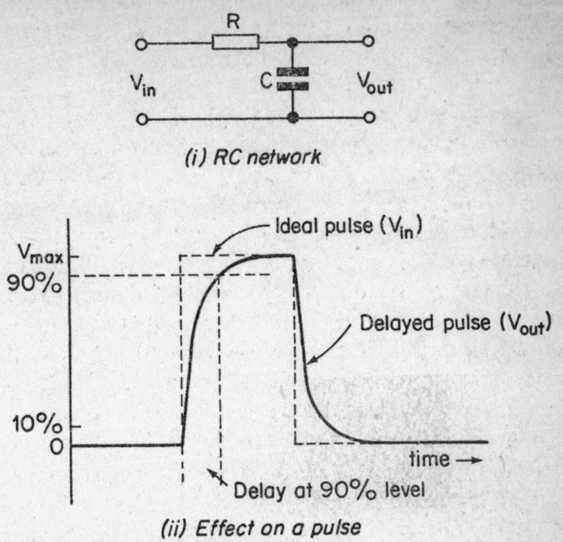

Fig. 5.16 Pulse delay with capacitance

5.5.1.2 Bandwidth

Although the requirement of large bandwidth for successful transmission of a pulse has been mentioned several times, Fig. 5.17 is now added not only to illustrate this graphically but also to show how delay can occur. It is only necessary to show half the rectangular pulse because the remainder is simply a mirror-image. The pulse is one of a train where T = 2t [Fig. 5.15(i)] so that each pulse is effective over the first 180° of the cycle. The various curves show a pulse after transmission over a circuit of limited bandwidth such that all harmonics above a certain one are excluded, for example, the curve marked 7 has all harmonics above the 7th removed. Thus if the pulse repetition frequency is 1000 Hz, frequencies above 7000 Hz are excluded. The graphs show clearly how the rise time decreases with increasing bandwidth, the decay

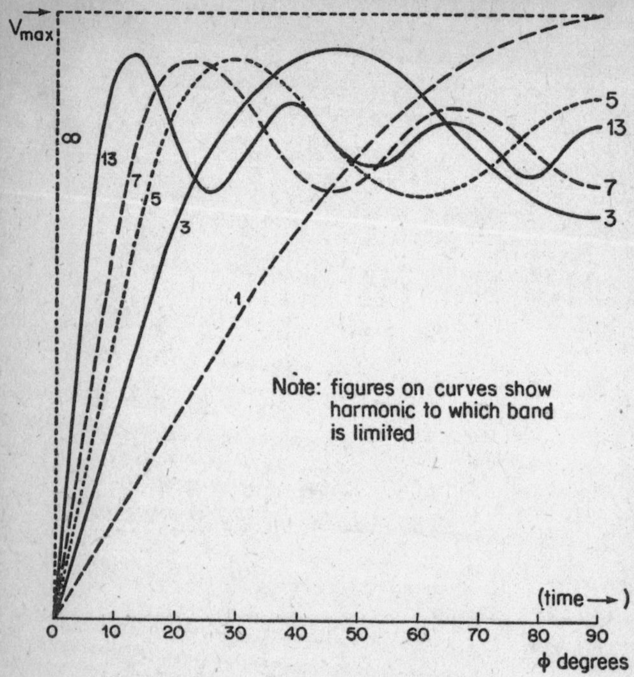

Fig. 5.17 Pulse shape with bandwidth limitation

time obviously decreases equally. The two extremes are also of interest, with no harmonics transmitted the pulse emerges as a sine wave, with an infinite number it would be received unscathed.

5.5.2 Pulse Modulation Techniques

Although earlier we considered p.c.m. almost exclusively and continue to do so after this section, this is not to say that other pulse-modulation systems are unimportant. Just as a sine-wave carrier can be modulated in several different ways, so too a pulse train can be modified. The various methods are set out in Fig. 5.18 showing the effects on a pulse-train (ii) of

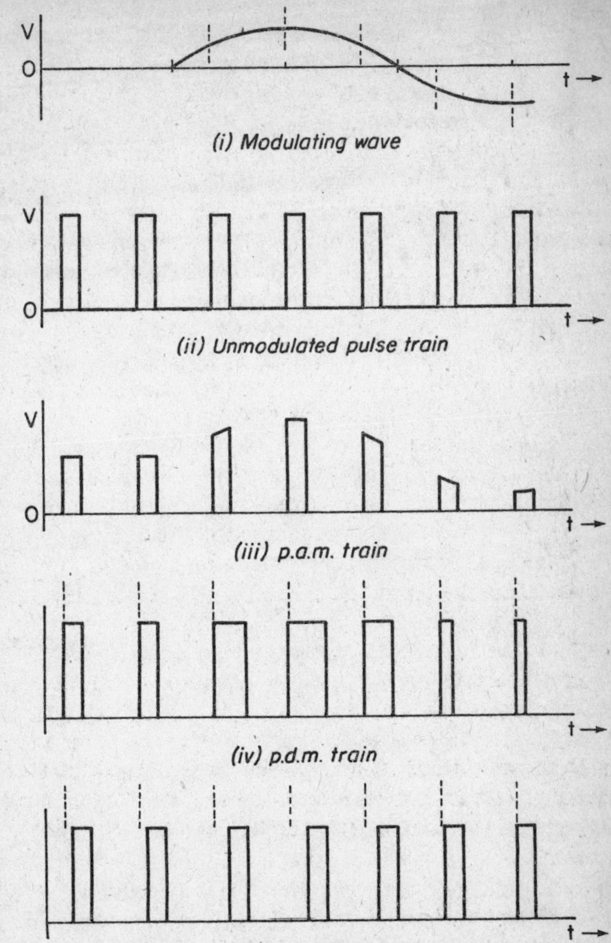

[Note: dotted lines indicate leading edges of pulses in (ii)]

Fig. 5.18 Pulse modulation

the modulating wave at (i).

Pulse Amplitude Modulation (p.a.m.) is shown in (iii) and it is evident that the amplitude of each pulse is according to that of the modulating wave at the instant of sampling. The fact that so few pulses are needed to sense the information content of the modulating signal (about one for each half-cycle) indicates that provided that the pulse length is kept short, ample time is left between pulses for those from other channels to be inserted, that is, p.a.m. is easily multiplexed.

Pulse Duration Modulation (p.d.m.), also known as pulse-width modulation, in (iv) employs pulses, the time duration of which accord with the modulation. The modulated pulse commences at the same time as its unmodulated counterpart but its trailing edge is retarded in time for positive excursions of the modulating wave and advanced in time for the negative excursions. The figure shows how the duration (or width) therefore varies.

Pulse Position Modulation (p.p.m.) in (v) has pulses shifting from their normal time positions. In the system shown, at maximum negative modulation the leading edge of the pulse occurs at the same time as in (ii), as the modulating signal moves positively, delay in the leading edge increases, thus with no modulation the p.p.m. pulses occur with some delay.

There are many variations of the three systems described, the fourth, p.c.m. is an ingenious development of p.a.m., the latter when transmitted over lines being more affected by interference voltages. Also compared with p.c.m. both p.d.m. and p.p.m. suffer more from pulse distortion caused by cables. Such a difficulty is inherent in any pulse system when pulses have very short time intervals between them. From Figs. 5.16 and 4.21(ii) it is evident that if a pulse overlaps the next digit position there is the possibility of error. From Fig. 5.18(iv) and (v) it can be seen that with varying gaps between pulses, when the gap decreases from the average, adjacent pulses may overlap more easily than for p.c.m. which maintains a constant time interval. When trouble occurs it is known as *intersymbol interference*.

5.5.3 Pulse - Code Modulation

Before we examine p.c.m. multiplex systems in detail it may be profitable to glance again at Fig. 4.22. The incoming speech is sampled at fixed intervals with an output p.a.m. train as in Fig. 5.18(iii). The encoder expresses each p.a.m. pulse in a code form which in fact is the p.c.m. signal. At the receiving end the p.a.m. signal is reconstituted, distributed to its channel equipment and demodulated to regain the original speech signal. In the 32-channel system we have chosen for study, time-slot 0 is used for frame synchronization and time-slot 16 for signalling information so leaving 30 slots for speech channels. It is therefore effectively a 30-channel system but needing 32 time slots. In some alternative systems all time slots carry a speech channel with synchronization and signalling effected by using 1 digit of each slot. In this case a similar 2.048 Mb/s system would have 32 speech channels, each with 7 digits available for speech transmission.

5.5.3.1 Sampling

Sampling is simply measuring the amplitude of a signal at certain (usually regular) time intervals. To obtain a single value for the sample it should be measured over a very short period of time so that amplitude changes are insignificant. In Fig. 5.18(iii) therefore the p.a.m. samples shown are not in accordance with this for the sketch shows some pulses with amplitudes varying between leading and trailing edges, in practice such pulse durations are excessive.

Now any p.c.m. system is limited mainly by bandwidth considerations in the number of bits it can have in each pulse code. Our practical system works to 8 bits and for this number there can be $2^8 = 256$ different codes.(4/A1.2.2) Effectively this means that the amplitude range of the modulating signal cannot be sampled precisely but only to the nearest of the 256 levels, rather like rounding numbers or money up or down. This is in fact working in steps and representing a signal by steps or discrete levels in this way is called *quantizing*. To compress 256 levels into a diagram detracts from clarity hence Fig. 5.19 shows the elements of a less practical system of only 4 digits and 5 channels.

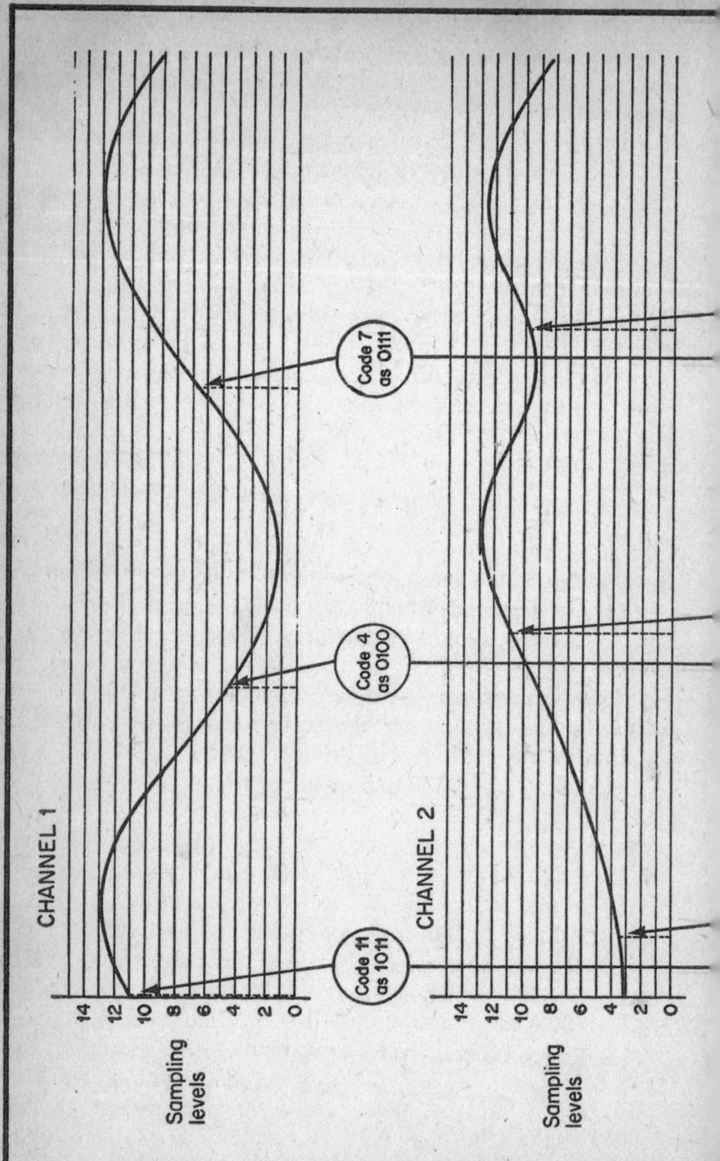
CHANNEL 1
CHANNEL 2
Code 11 as 1011
Code 4 as 0100
Code 7 as 0111
Sampling levels
14
12
10
8
6
4
2
0
Sampling levels
14
12
10
8
6
4
2
0

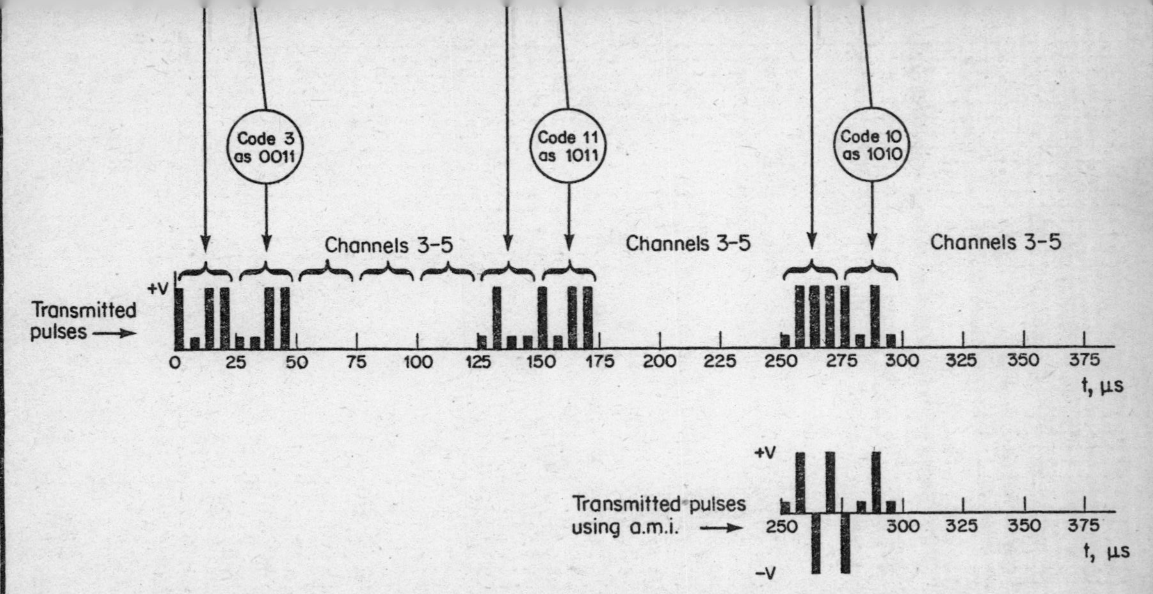

Fig. 5.19 PCM sampling and coding

The input modulation to Channel 1 here is a sine wave of 3.4 kHz. 4 digits allows $2^4 = 16$ sampling levels and it is assumed Channel 1 is sampled at t = 0, 125, 250 . . .μs (i.e. at 8 kHz). At t = 0 the nearest sampling level is 11 (say, mV). Any value between just over 10.5 and 11.5 is interpreted as 11 and the inaccuracy of quantizing begins to show. It is known as the *quantizing error*. The practical system has 256 levels instead of 16, hence the accuracy is 16 times better. The Channel 1 sample taken at 125μs has 4 as the nearest level and again at 250μs it is 7. Channel 2 in the figure is sampled at 25, 150 and 275μs resulting in sampling levels of 3, 11 and 10 respectively.

There must obviously be some impairment from the use of discrete sampling levels otherwise even fewer could be used. Consider Fig. 5.20(i) which shows the input/output characteristic of a quantized system as for Fig. 5.19. The ideal characteristic is a straight line such that whatever the amplitude of the input, that becomes the pulse amplitude at the output. The practical characteristic has a staircase form, for example, for all inputs between just over 0.5 and 1.5 the output is consistently 1. We can look at the discrepancy between practical and ideal a little more closely as follows:

Input	Output	Error
0	0	0
0.25	0	–0.25
0.5	0	–0.5
0.51	1.0	+0.49
0.75	1.0	+0.25
1.0	1.0	0
1.25	1.0	–0.25
1.5	1.0	–0.5
1.51	2.0	+0.49
1.75	2.0	+0.25
⋮	⋮	⋮

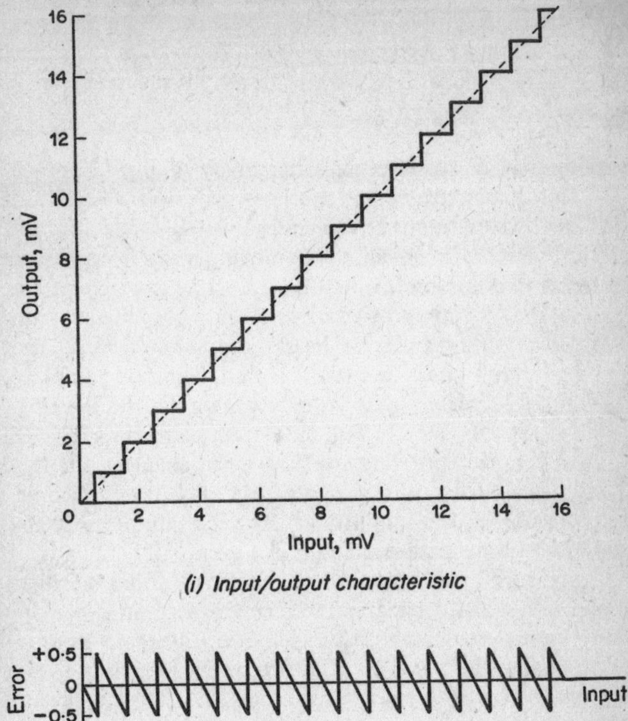

(i) Input/output characteristic

(i) Error

Fig. 5.20 Quantizing error

and then draw the graph of quantizing error as in Fig. 5.20(ii).

Effectively therefore the sampler output can be considered as a series of p.a.m. pulses which when reconstructed are equivalent to ideal pulses disfigured by the addition of an error signal. Our graph shows the latter to be of triangular form and Fourier will tell us that this abounds with harmonics, the result being that quantizing error gives rise to a system noise and it is usually referred to as *quantizing noise*. This noise is generated by the p.c.m. system itself, there is always the problem of unwanted noise being picked up, this is a system which generates its own! Thus we see the need to employ as many coding digits per sample as possible for then the steps in Fig. 5.20(i) and accordingly the error amplitude in (ii) become smaller.

Sampling may be made more efficient by taking advantage of the fact that in speech and music, signals of low amplitude occur much more frequently than do those of large amplitude. There is therefore an advantage in moving away from uniform quantizing to a non-uniform method. As one example of many the signal may be compressed in amplitude in such a way that the larger amplitudes receive most compression while the very small amplitudes are little affected. Then on sampling, relatively more of the quantizing levels act on the lower amplitudes as suggested in Fig. 5.21(i) on comparing (a) with (c). Although the sampling levels have not changed, the signal has and in such a way that progressively more levels become effective at the lower amplitudes. By such compression the signal has become distorted and if not eventually expanded to its original form, then speech or music would sound peculiarly flat and unnatural. Hence at the receiving-end an *expander* having the inverse characteristic to that of the *compressor* is used to restore the signal and the two devices together form a *compander*. This is an example in which the sampling levels are unchanged, in alternative methods more levels may be provided for the more frequently occurring lower signal amplitudes with progressively less for higher. This is illustrated by Fig. 5.21(ii) which shows both low and high-level signals and how they are affected by quantizing steps which increase with sampling level. Generally, with 256 sampling levels and non-uniform quantizing, quantization noise on telephony p.c.m. circuits is unlikely to be noticed by most users, although with fewer levels it is. However, when systems follow each other in a long-distance connexion, noise generated by any one in the chain is accepted as part of the signal by the next. Thus it is when several systems are connected in tandem that quantization noise becomes troublesome. [256 sampling levels is clearly different from the few we worked with in

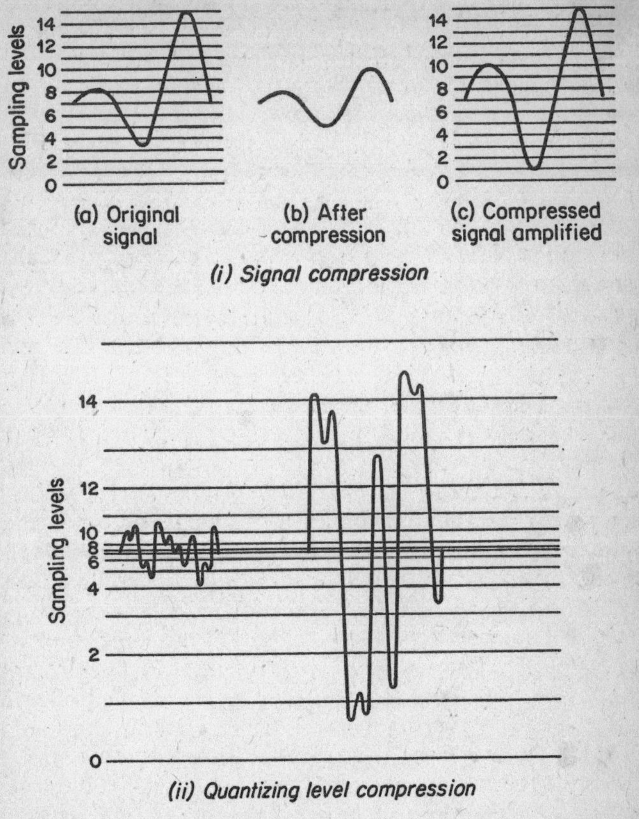

Fig. 5.21 Non-uniform quantizing

Chapter 1 when discussing information theory, but note that here we are considering the reduction of quantization noise, not the amount of information carried].

Electronic switches are invariably employed for sampling. Fig. 5.22(i) shows an elementary arrangement with one of the least complicated types of switch, a bridge of four diodes. The

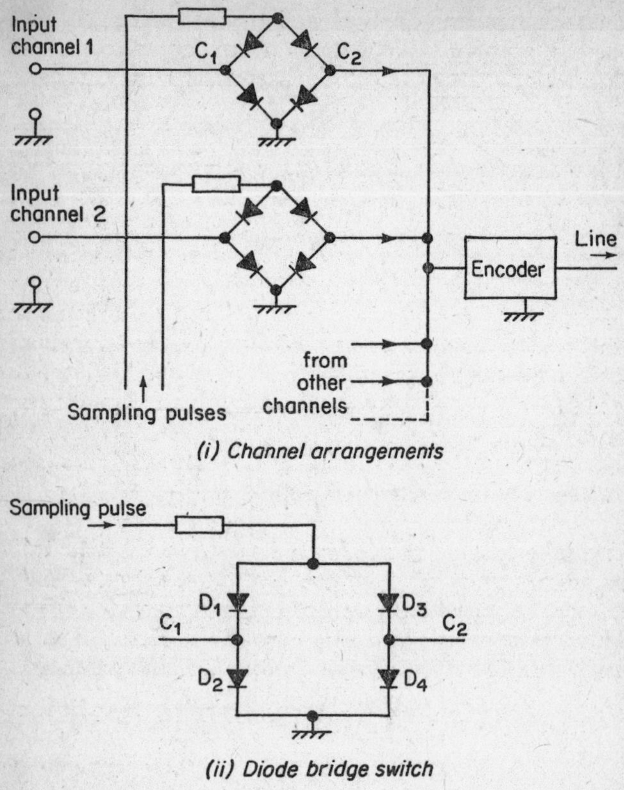

Fig. 5.22 Sampling

bridge is re-drawn in (ii) to show that with a positive sampling pulse, diodes D_1 and D_2 in series, also D_3 and D_4 in series are forward biassed, hence of low resistance. As far as the channel connexions are concerned, between points C_1 and C_2 are connected D_1 and D_3 in parallel with D_2 and D_4. With the diodes all at low resistance, C_1 is virtually switched through to C_2, so the channel input is connected to the encoder. The sampling pulses are supplied from the system clock source, they are of short duration and are connected to each channel in turn (see Fig. 4.21 for the mechanical equivalent). When a sampling pulse is not present on a channel, a negative or no bias exists with the diodes therefore at high resistance and the channel input effectively disconnected from the encoder.

5.5.3.2 Coding

As far as the communication engineer is concerned, a code is a set of rules developed for conversion of a message or information from one form into another. A well-known example on which the computer world rests is the binary code in which information in the form of one set of numbers (decimal) or characters is converted into a different set (binary). Any code can be used for a p.c.m. system, binary was used in the early days and because it is still much in use and so well known to computer people, we take p.c.m. coding into binary as our example. What is required is no more than that a sampling level number (say, any from 0 to 255) is converted into the appropriate binary number.(4/A.1)

Each digit in a binary number has a decimal value according to its position. Whether this value is added in as part of the whole number or not depends on the binary indication of 0 or 1. This is most easily illustrated by examples, firstly as shown in Fig. 5.19, then for an 8 digit system. Decimal values are:

binary digit →	4th	3rd	2nd	1st(least significant)
	2^3	2^2	2^1	2^0
decimal value	8	4	2	1

and hence the decimal value of binary 1011 is

$$(1 \times 2^3)+(0 \times 2^2)+(1 \times 2^1)+(1 \times 2^0) = 8+0+2+1 = 11$$

and 0111

$$(0 \times 2^3)+(1 \times 2^2)+(1 \times 2^1)+(1 \times 2^0) = 0+4+2+1 = 7$$

the transmitted pulses which carry these binary numbers are shown in the figure. Moving on to the modern 8-digit system, identical principles apply, for example 98 is expressed as 01100010 and 255 as 11111111.

There are certain disadvantages when transmitting pulse signals in the form shown in Fig. 5.19 over lines which contain transformers (most do). If all pulses (1's) are in one direction only (positive in the figure) there is a d.c. component and this is not transmitted by a transformer. Accordingly pulses reach different values depending on the signals immediately preceding them especially with long strings of 0's or of 1's. A modification of the system is to use a *ternary* or 3-level code with the idea of eliminating a d.c. component. This is accomplished by alternately inverting the 1's between positive and negative as shown at the bottom of the figure. The coding system is known as *bipolar* or as *alternate mark inversion* (a.m.i.), "mark" being the telegraphy way of expressing a 1 (Section 4.3.1).

5.5.3.3 Modulating Signal Reconstruction

The p.c.m. line signal, a series of 0's and 1's as shown in Fig. 4.21(ii) after regeneration meets a decoder. The stream of binary digits is first divided into 8-digit groups, correct grouping being maintained by reception of some type of synchronizing signal from the sending end. Decoding of each group then proceeds as soon as the 8th digit is received. The decoder is a digital/analogue converter,(4/4.8.1) of which there are many types available, its output being a replica of the sending-end quantized p.a.m. samples, one pulse output for each 8-digit p.c.m. group input.

The distributor is controlled by clock pulses which open gates (i.e. switch) at the appropriate times so that each channel demodulator receives its correct stream of p.a.m. pulses (at 125μs intervals). The channel demodulators are simply low-pass filters having cut-off frequencies just above the highest modulating frequency. To understand this in detail involves a lot of mathematics but since we only need an appreciation of what happens we can get by with a little. What follows therefore is a simplified approach based mainly on the analysis methods which we are accustomed to using.

Because it is of rectangular form, the sampling pulse train can be considered as a series of harmonics of the pulse-repetition frequency, f_c, plus a d.c. component arising because the pulse train is unidirectional. An equation to the pulse train expresses the instantaneous value, ν_c as

$$\nu_c = V_{dc} + a_1 \sin \omega_c t + a_2 \sin 2\omega_c t + a_3 \sin 3\omega_c t + \ldots$$

V_{dc} is of course the d.c. component and a_1, a_2, a_3 etc the amplitudes of the various harmonics, $\omega_c = 2\pi f_c$. Now the modulating frequency is given by $\nu_m = V_m \sin \omega_m t$ and for convenience as before, we consider one modulating frequency f_m only and expand it later to the full band.

The pulse train is amplitude modulated by the modulating signal f_m. As we saw in Section 5.2, mathematically the two equations are multiplied together which gives:

$$\nu_{pam} = (V_{dc} \cdot V_m) \sin \omega_m t + a_1 V_m \sin \omega_c t . \sin \omega_m t$$

$$+ \; a_2 V_m \sin 2\omega_c t . \sin \omega_m t + \ldots .$$

(for convenience we leave out any higher harmonics because we shall find below that they are filtered out but the + . . . reminds us of their existence). Then

$$\nu_{pam} = (V_{dc} \cdot V_m) \sin \omega_m t$$

$$+ \frac{a_1 V_m}{2} \left[\cos (\omega_c - \omega_m)t - \cos (\omega_c + \omega_m)t \right]$$

$$+ \frac{a_2 V_m}{2} \left[\cos (2\omega_c - \omega_m)t - \cos (2\omega_c + \omega_m)t \right]$$

$$+ \ldots .$$

(by use of equation (v) of Appendix A3.2). In terms of frequency instead of angular velocity:

$$\nu_{pam} = (V_{dc}\colon V_m)\sin 2\pi f_m t$$
$$+ \frac{a_1 V_m}{2}\left[\cos 2\pi(f_c - f_m)t - \cos 2\pi(f_c + f_m)t\right]$$
$$+ \frac{a_2 V_m}{2}\left[\cos 2\pi(2f_c - f_m)t - \cos 2\pi(2f_c + f_m)t\right]$$
$$+ \ldots.$$

Thus one of the components is the modulating frequency, the remainder are lower and upper sidebands on each carrier harmonic, theoretically right up to infinity. Note that the carrier itself is excluded.

Now what we are aiming for is to find how we can isolate the component containing f_m on its own, for this is demodulation, that is, sorting out the modulating frequency from the p.a.m. carrier. Let us see the expression as a frequency diagram as in Fig. 5.23(i) where we now expand f_m into a *band* of frequencies, say, 300 – 3400 Hz. The first component of the expression above, being the modulating frequency, extends over this range as shown. The second component containing $(f_c - f_m)$ and $(f_c + f_m)$ extends for $f_c = 8$ kHz in two sidebands from 4.6 to 7.7 kHz and 8.3 – 11.4 kHz. The third has sidebands on 16 kHz, part of the lower is shown. The figure shows clearly how the modulating frequency band is extracted, simply by passing the p.a.m. train through a low-pass filter cutting off between 3.4 and 4.6 kHz, everything above this being rejected.

We also begin to understand the requirement of sampling, in this case at 8 kHz, or theoretically at $2f_{m\ max}$. If sampling were at the minimum value, here at 6.8 kHz, it is clear that the modulating frequency band and the first lower sideband would coincide at 3.4 kHz as shown in Fig. 5.23(ii) and no filter could satisfactorily separate them. With a sampling frequency

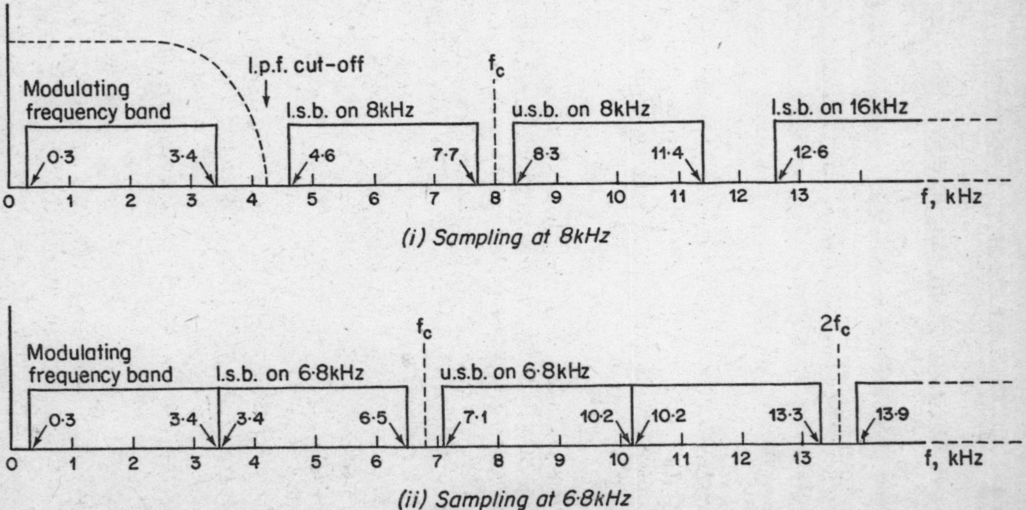

Fig. 5.23 Frequency spectra of p.a.m. pulse trains

less than $2f_{m\ max}$, the two bands overlap, making perfect demodulation impossible.

We have only got to grips with sampling at the lower audio frequencies as used in telephony. For the higher frequencies of television the theory is unchanged, thus a sampling frequency of at least 11 MHz is required for a video colour signal.

6. THE ELECTROMAGNETIC WAVE IN COMMUNICATION

A circuit which gives us no difficulty is shown in Fig. 6.1(i), an a.c. generator (of speech, pure tones etc) connected to a load resistance R_L. We have always understood that the instantaneous value of the current is the same anywhere in the circuit for if it is not we have deceived ourselves when using Ohm's Law. But we need not be too concerned, provided that the generator and load are reasonably close together or the frequency not too high, all these ideas hold good. It is when the wires which connect the generator to the load leave the confines of the equipment case that they form a *transmission line* [Fig. 6.1(ii)] and we become involved in a whole new field of study in which electrical energy moves from one point to another in a form of wave motion. The same occurs in radio transmission. What is happening is that "wave" has until now conjured up something on a graph with time on the X-axis (or radians or degrees, but these are also related to time), in this Chapter we shall be more concerned with visualizing waves, not how they vary with time at a certain point but how

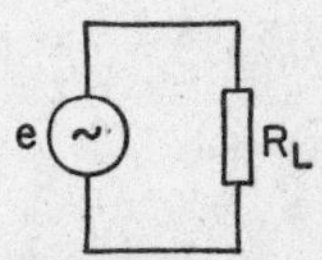

(i) Generator with nearby load

(ii) Generator with remote load

Fig. 6.1 Connecting a generator to a load

voltage and current change at different distances along a transmission path. It is preferable to start with material lines as opposed to radio paths because in a way these seem more tangible.

6.1 TWO-WIRE TRANSMISSION LINES

We firstly look at a few of the things which happen when pair length becomes significant:

(i) from the generator's point of view the resistance of the pair of wires is added to R_L, for long pair lengths it may be greater than R_L, in fact many times

(ii) current leakage through the insulation separating the wires while ineffectuál over a short distance becomes appreciable for long distances

(iii) the pair has capacitance, it has of course in (i) of the figure but there is it usually negligible.

(iv) the earlier discussion on skin effect (Section 1.6.2.2) reminds us that even a straight wire has inductance, this adds up as length increases

(v) although very small in human experience, time is needed for a current to travel from the generator to the load. This brings us to the point about wave motion.

It is perhaps quite easy to form a picture in the mind of electrons jostling along a wire in the spaces between the atoms but when we expand this to visualize also the fact that energy travels along lines (or through the atmosphere) as "waves", the subject does suddenly seem to become complicated. We need not be dispirited for although we have not the mathematics at our disposal for an exhaustive study, we can at least make a start, not only to be convinced that it does happen, but also to become acquainted with just a few of the elementary principles. Let us first see pictorially how it is that a wave "travels" along a transmission line.

6.1.1 Wave Motion

It is often more illustrative if instead of using symbols (x, y, t etc), we work in practical figures. Those used here are chosen for convenience, any others are equally applicable. Fig. 6.2 shows what happens along a 2-wire transmission line 3km long from the moment a sinusoidal voltage $V\cos\omega t$ is applied at the sending-end. We use a cosine wave (which after all is only a sine wave starting earlier) so that at the onset a voltage is applied to the line rather than nothing. The frequency is 100 kHz and assuming that a current flows down the line at the speed of light (which is approximately true for open-wire lines but certainly not for cables, however this in no way affects the explanation) then the wavelength

$$\lambda = \frac{c}{f} = \frac{3 \times 10^8}{10^5} = 3000\text{m}$$

and the period (time for 1 cycle)

$$= \frac{1}{f} = \frac{10^6}{10^5} = 10\mu\text{s}.$$

With these figures Fig. 6.2 can be labelled. Assume V = 1 volt.

In (i) of the figure is the transmission line running from 0 to 3km, (ii) shows the graph of the sine wave voltage applied at the sending-end. We look at the voltages produced by the current in the line at (iii) as time progresses.

At t = 0 the sending-end voltage $V_s = 1 . \cos \omega t = 1 . \cos (2\pi \times 10^5 \times 0) = 1.0$V. This is marked above the line representing the transmission line in (i) by a small arrow, its distance away from the line indicating the value of the voltage at that point.

At $t = 1\mu s$ V_S has fallen to $1 . \cos (2\pi \times 10^5 \times 1 \times 10^{-6}) = 0.81$V. However the full 1V applied 1μs earlier has now progressed down the line $3 \times 10^8 \times 10^{-6}$m (c x t) = 300m, this is shown on the next line down for t = 1μs.

Thus we can plot the voltage at any point along the line as is done in the figure for t = 2.5, 5.0, 7.5 and 10μs by noting the progress of the original applied voltage at distances along the line while also calculating V_S. Note that we have taken no account of attenuation which causes the line voltage to decrease in magnitude as it progresses, this in no way affects the present intention of seeing how a wave is set up.

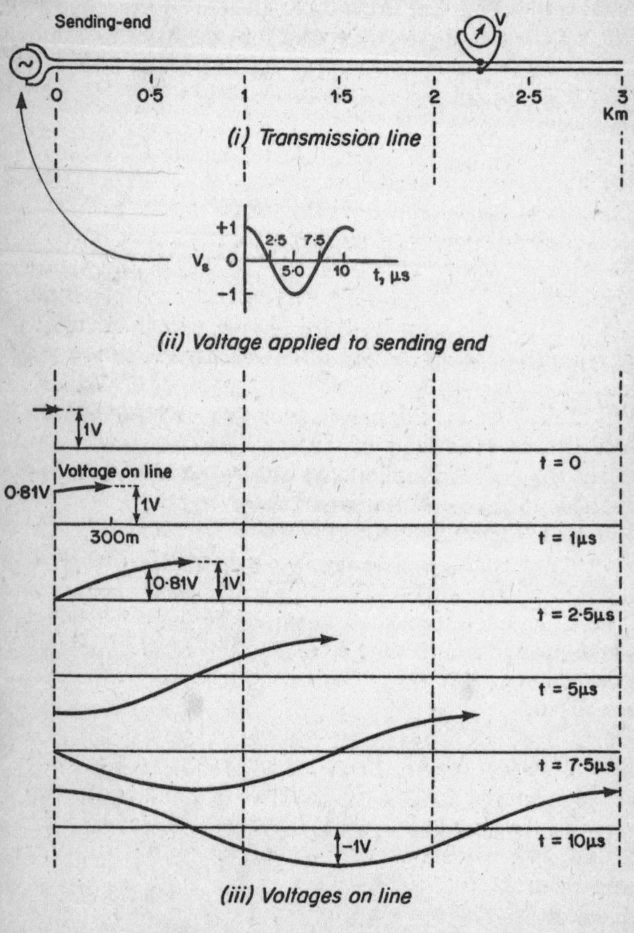

Fig. 6.2 Progression of a voltage wave

Let us concentrate on the findings at t = 10μs for after this point the process repeats and the pattern is unchanged because V_s has gone through one complete cycle. A voltmeter connected for example across the line at 2.25 km from the sending-end [see (i)] would measure nothing, yet at 1.5km it would show −1V. 2.5μs earlier the values were +1V and 0. We would expect the values to rise and fall sinusoidally but hardly to vary significantly with distance. [We shall be able to compare this later with radio transmissions where again the voltage at a point in space rises and falls as the radiated wave travels past].

If the line is more than 3 km long, the wave simply progresses to the end, continually falling in amplitude because of the line losses. However, just suppose it is open-circuit at 3 km, i.e. there is no R_L, what then? Looking again at the drawing for t = 10μs there is energy being continuously applied at the sending-end, flowing down the line but suddenly coming to a halt. A current cannot be forced through a load, it must go somewhere and the only way is back, that is, it is *reflected* and returns to the sending-end. On its way it adds to or subtracts from the originating wave and clearly causes havoc when an attempt is made to measure line voltage. This phenomenon has many parallels in daily life. If, for example a wave is formed on water in a bath with the flat of the hand so that it travels towards the vertical end, the reflexion is quite evident.

Voltage on a line gives an electric field, the current flow produces a magnetic flux, hence an *electromagnetic wave* is transmitted.

So far this is just food for thought and mainly to show what lies ahead in this somewhat different but not conflicting approach. Remember none of the rules has really been changed, we are now only seeing things on *both* a time and a distance basis.

6.1.2 Line Coefficients

Various *coefficients* are needed to give a full electrical description of a transmission line so that the influence it has on an electromagnetic wave can be determined. Some are brought to light in (i) to (iv) of Section 6.1 and we now look at these *primary coefficients* in greater detail, followed by some *secondary coefficients* which arise from them. By knowing the values of the coefficients for any particular line, attenuation, wave velocity and phase relationships at any frequency can be calculated.

6.1.2.1 Primary Coefficients

Points (i) to (iv) of Section 6.1 can be summed up by saying that a transmission line has four primary coefficients:

R, the resistance in ohms G, the leakance in siemens C, the capacitance in farads L, the inductance in henries	or in submultiplies of the units per unit length of line (usually 1 metre or 1 mile).

The resistance is that of the two wires (*loop resistance*) over the unit length and includes skin effect. As we saw in Section 1.6.2.2, at the higher frequencies the effect varies as $\sqrt{f}$ so that the actual resistance may be up to many times the d.c. resistance.

Leakance is a new term, it is the effect or loss due to the conductance[1/3.1] of the insulating material between the wires. Like conductance it is measured in siemens. The leakage current is frequency dependent because not only does it have the normal non-reactive component flowing through the insulation but also a component due to the power losses in the dielectric when the line capacitance is charged and discharged. The two components together give the total leakance G.

Capacitance in a 2-wire line has already been noted for its undesirable shunting effect as frequency increases. It is less pronounced for overhead lines on poles owing to the greater wire spacing than for underground cables. There is little change with frequency.

Inductance is that for the two-wire loop over the unit length.

There is also mutual inductance(1/5.3.2) between the wires and because the current in them at any instant is in opposite directions, the total inductance is given by $L_1 + L_2 - 2M$ where L_1 and L_2 are the self-inductances of the wires and M the mutual inductance between them. From this, the greater the spacing between the wires (lower M), the higher the inductance becomes. Inductance does not vary appreciably with frequency.

Some practical figures are given in Table 6.1 for two very different types of transmission line, the first on underground cable with 1.3mm diameter paper-insulated copper wires and the second a less frequently encountered overhead 2.6mm copper pair carried 20 cm apart. The figures are per loop mile.

TABLE 6.1 VALUES OF PRIMARY COEFFICIENTS FOR TWO TYPICAL TRANSMISSION LINES

Transmission Line	frequency kHz	R Ω	G µS		C µF	L mH
1.3mm in Underground Cable	10	44	18		0.06	1.1
	100	90	400		0.06	0.96
			Dry*	Wet*		
2.6mm Overhead	10	13	1	6.5	0.009	3.4
	100	35	9	27	0.009	3.3

*very much dependent on weather and type of insulators used.

The effect of frequency on both R and G is evident, equally the little effect it has on C and L. The rise in inductance and fall in capacity with increase of wire separation is also evident.

6.1.2.2 *Characteristic Impedance*

We next use the primary coefficients to develop secondary ones, the *characteristic impedance* and the *propagation constant*. The first is important because when a line is terminated in its characteristic impedance, it is then matched to its termination and delivers maximum power to it (Section

3.1.4), with no energy left over to be reflected. From the propagation constant we obtain the line attenuation, wave velocity and phase relationships.

By definition the characteristic impedance is that measured at the sending-end of an infinitely long line, so with such an impossible length it would only be necessary to apply a voltage V to one end and measure the current I to obtain the input or characteristic impedance Z_0 from V/I as shown in Fig. 6.3(i) But suppose a short length is cut off the near-end of the line as in (ii) so creating terminals 3, 4, 5 and 6. Looking down the line from terminals 5 and 6 the impedance is still Z_0 because

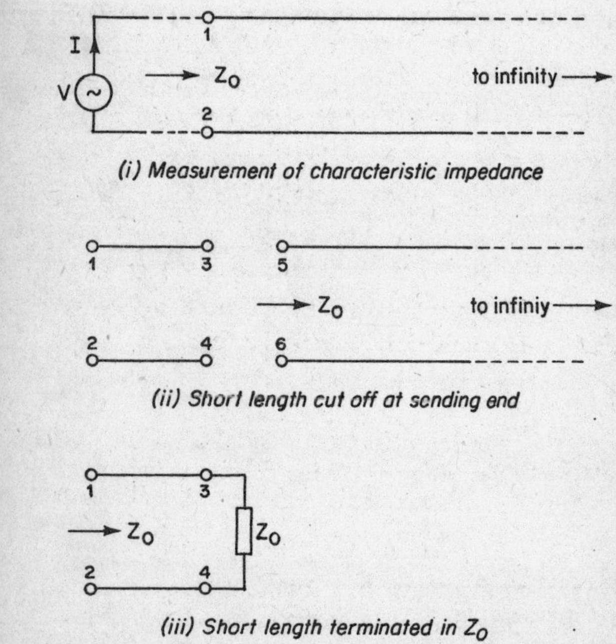

Fig. 6.3 Characteristic impedance

the infinite nature of the line is unaffected, therefore by connecting an impedance Z_0 to terminals 3 and 4 as in (iii), the impedance looking into terminals 1 and 2 is identical with that in (i), that is, the short length of transmission line in (iii) behaves as thought it were infinitely long. This makes some sense except that we could not have had an infinite line on which to measure Z_0 in the first place, fortunately suitable measurements can be made on the short line as we will see later.

The primary coefficients are distributed, meaning that the wave does not experience the whole resistance, capacitance etc. in one lump, it meets them bit by bit along the length of the line. Therefore we cannot represent the line as a single 4-terminal network, unless it refers to a very short length, so

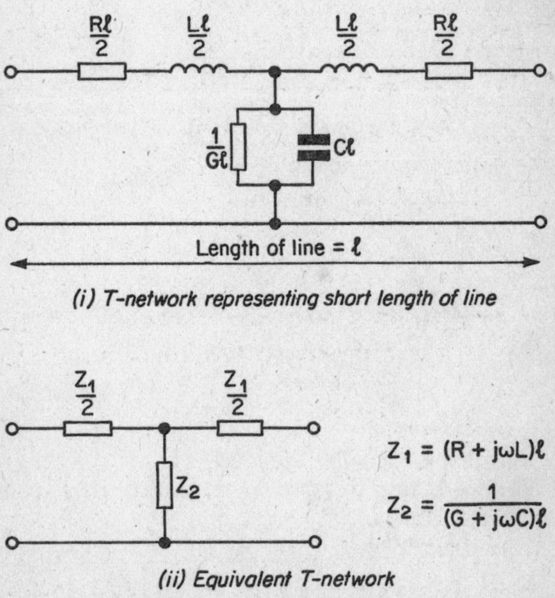

Fig. 6.4 Simulation of line by T-networks

short in fact that the coefficients can reasonably be considered as lumped. Nevertheless let us start by considering a section of a line of length, say, ℓ metres or miles etc (according to the units in which the primary coefficients are quoted) and represent it by the equivalent T-network as in Fig. 6.4(i). Both resistance and inductance may be considered as being in series while the capacity is naturally connected across the wires. The leakance is in parallel with C, changed to a resistance having a value 1/G ohms (resistance is the reciprocal of conductance). We can look at this network on a generalized impedance basis as in (ii) where $Z_1 = (R + j\omega L)\ell$ and Z_2 is the resultant of

$$\frac{1}{G\ell} \text{ in parallel with } \frac{1}{j\omega C\ell} \quad (1/3.4.5) \quad \text{i.e.}$$

$$Z_2 = \frac{\dfrac{1}{G\ell} \times \dfrac{1}{j\omega C\ell}}{\dfrac{1}{G\ell} + \dfrac{1}{j\omega C\ell}} = \frac{\dfrac{1}{j\omega CG\ell^2}}{\dfrac{(G + j\omega C)\ell}{j\omega CG\ell^2}} = \frac{1}{(G + j\omega C)\ell}$$

Now as shown in Section 3.2.3.1, Z_0 for the T-network in Fig. 6.4(ii) when terminated in Z_0 is

$$Z_0 = \sqrt{\frac{Z_1{}^2}{4} + Z_1 Z_2}$$

(R_1 in the earlier section becomes $Z_1/2$ here and R_2 becomes Z_2)

$$\therefore\ Z_0 = \sqrt{\frac{(R + j\omega L)^2 \ell^2}{4} + \frac{R + j\omega L}{G + j\omega C}}\ .$$

Ultimately when the length ℓ approaches 0, the characteristics of the T network tend to represent accurately those of the line. When this happens

$$\frac{(R + j\omega L)^2 \ell^2}{4} \text{ approaches 0 and}$$

$$Z_0 = \sqrt{\frac{R + j\omega L}{G + j\omega C}}$$, a useful and important equation

For practice we calculate the characteristic impedance of the 1.3mm underground cable pair of Table 6.1 at 10kHz.

$R = 44\Omega$, $L = 1.1$ mH, $G = 18\mu S$, $C = 0.06\mu F$ (per loop mile).

$$\text{Then } Z_0 = \sqrt{\frac{R + j\omega L}{G + j\omega C}}$$

$$= \sqrt{\frac{44 \text{ x } j2\pi \text{ x } 10^4 \text{ x } 1.1 \text{ x } 10^{-3}}{18 \text{ x } 10^{-6} + j2\pi \text{ x } 10^4 \text{ x } 0.06 \text{ x } 10^{-6}}}$$

$$= 10^3 \sqrt{\frac{44 + j69}{18 + j3770}}$$

$$= 10^3 \sqrt{\frac{81.8\angle 57.5^\circ}{3770\angle 89.7^\circ}}$$

[remember:

division of polar form: $p\angle\theta \div r\angle\phi = \frac{p}{r}\angle\theta - \phi$

and square root : $\sqrt{r\angle\phi} = \sqrt{r}\angle\frac{\phi}{2}$](2/1.3.4)

$\therefore\ Z_0 \approx 147\angle -16^\circ$ or $(140 - j40)$ ohms.

It can also be shown by further consideration of the T-network that if Z_{oc} is the impedance measured looking into a practical line with its far-end disconnected and Z_{sc} that when the far-end is short-circuited, then $Z_0 = \sqrt{Z_{oc} . Z_{sc}}$ thus giving a convenient means of measuring Z_0 in the field.

Generally both cable and overhead pairs have values of Z_0 of the order of several hundred ohms.

6.1.2.3 *Propagation Coefficient*

This coefficient we denote by p (P and γ, Greek "gamma" are also used). It is expressed in nepers which can, of course be converted into decibels, hence from Section 2.1.1.3:

$$p = \log_e \frac{I_1}{I_2}$$

where I_1 and I_2 are the currents at two different points on a transmission line. It is equally possible to work in voltages. p is a vector quantity because the two currents are not in phase owing to the time taken for the wave to travel between the two points. Imagine a line to be split up into its elementary equivalent T-networks as in Fig. 6.5, the currents in the various sections being labelled as shown. We wish to determine the propagation coefficient per section, hence of the line. Then since

$$p = \log_e \frac{I_s}{I_1} \text{ (first section) } e^p = \frac{I_s}{I_1} \qquad (2/A4.2.1)$$

Also for the other sections

$$e^p = \frac{I_1}{I_2} = \frac{I_2}{I_3} = \ldots .$$

$$\therefore \frac{I_s}{I_2} = \frac{I_s}{I_1} \times \frac{I_1}{I_2} = e^p \times e^p = e^{2p}$$

and

$$\frac{I_s}{I_n} = e^{np}$$

where n is the number of sections, so the value of the current leaving the nth section $I_n = I_s . e^{-np}$ and if the sections are each 1 metre in length and the primary coefficients are quoted per metre then n is the distance in metres. Similarly for miles.

This shows how we can calculate the current at any point provided that the propagation coefficient is known.

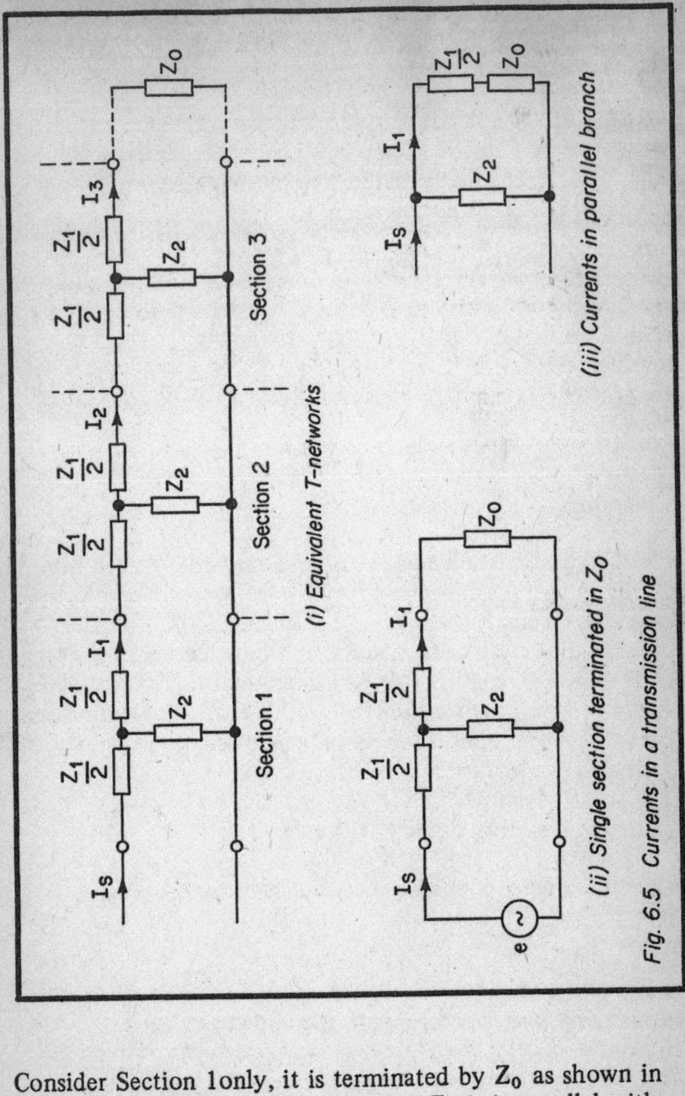

(i) Equivalent T-networks

(ii) Single section terminated in Z_0

(iii) Currents in parallel branch

Fig. 6.5 Currents in a transmission line

Consider Section 1only, it is terminated by Z_0 as shown in Fig. 6.5(ii) and from (iii) we see that Z_2 is in parallel with $(Z_1/2 + Z_0)$(1/3.4.5)

$$\therefore \frac{I_1}{I_s} = \frac{Z_2}{Z_2 + Z_1/2 + Z_0} = e^{-p}$$

Next the equivalent values of the primary coefficients for Z_1 and Z_2 are substituted [Fig. 6.4(ii)], but there is little point in going through the mathematics step by step a second time so we accept that the solution is

$$p = \sqrt{(R + j\omega L)(G + j\omega C)}$$

which, as in the case of Z_0, is a vector quantity. The quadrature symbols generally used are α and β such that $p = \alpha + j\beta$. Since p states the relationship between the input and output currents of a section, then its real part (α) gives the attenuation of the section in nepers for

$$\alpha = \log_e \left| \frac{I_1}{I_2} \right| \text{ nepers}$$

α is accordingly known as the *attenuation coefficient*. The imaginary part (β) indicates the phase difference between the two currents and it is known as the *phase coefficient*. Thus, knowing the length of the section and the phase change over it (β), the wavelength λ can be calculated for if the phase change is β radians over the section, there are $2\pi/\beta$ sections to one wavelength. Thus if the primary coefficients are quoted per metre, we let the section be 1 metre in length and

$$\lambda = \frac{2\pi}{\beta} \text{ metres, similarly for miles.}$$

Also the velocity of the wave at any frequency,

$$\nu = f\lambda = \frac{2\pi f}{\beta} = \frac{\omega}{\beta}$$

metres/s or miles/s according to the units used.

Transmission line theory in general and the propagation coefficient in particular can hardly be said to be easy to grasp especially when the latter is found to include a mixture of epsilon, phase differences and complex (j) notation. We therefore take an example, and go through the calculations, seeing what they tell us as we progress. The figures used are for a practical underground copper pair, the wires being 0.91 mm diameter and paper insulated. We work at a frequency of 1600 Hz where $\omega = 10{,}000$ rad/s which slightly reduces the arithmetic.

$R = 84.1\Omega$/mile, $L = 1.11$ mH/mile,

$G = 1.75\mu S$/mile, $C = 0.0608\mu F$/mile.

$$R + j\omega L = 84.1 + j10^4 \times 1.11 \times 10^{-3}$$

$$= 84.1 + j\,11.1 = 84.83 \angle 7.52^\circ$$

$$G + j\omega C = 1.75 \times 10^{-6} + j10^4 \times 0.0608 \times 10^{-6}$$

$$= 10^{-6}\,(1.75 + j608)$$

$$= 10^{-6}\,(608 \angle 89.84^\circ)$$

$$p = \sqrt{(R + j\omega L)(G + j\omega C)}$$

$$= 10^{-3}\sqrt{(84.83\angle 7.52^\circ)(608\angle 89.84^\circ)}$$

$$= 10^{-3}\sqrt{51577\angle 97.36^\circ} = 10^{-3}(227.1\angle 48.68^\circ)$$

$$= 0.227\angle 48.68^\circ$$

$$= 0.227\cos 48.68^\circ + j0.227\sin 48.68^\circ$$

$$= 0.15 + j0.17$$

i.e. $\alpha = 0.15$ nepers per mile and $\beta = 0.17$ radians per mile

α also equals $0.15 \times 8.686 = 1.3$dB/mile

From the phase coefficient we can calculate wavelength (λ) and wave velocity (ν).

$$\lambda = \frac{2\pi}{\beta} = \frac{2\pi}{0.17} \approx 37 \text{ miles}$$

$$\nu = \frac{\omega}{\beta} = \frac{10^4}{0.17} \approx 58{,}800 \text{ miles/second, about one-third the}$$

the speed of light. Thus the commonly held idea that all electromagnetic waves travel at the speed of light is misleading.

6.1.3 The Distortionless Condition

Having calculated the propagation coefficient for a particular cable at a single frequency as an exercise, we might next struggle through similar calculations for a range of frequencies. Some even more interesting features of the transmission line would then be revealed and all these are graphically shown in Fig. 6.6. The variations in α, β, ν and Z_0 are shown for frequencies from 1 to 100 kHz. We need not be too concerned with actual values because these refer to one typical cable only but what concerns us more is that none of the coefficients is constant with frequency. In (i) α and β both rise with frequency as might be expected since ω in the expression for p rises and added to this is also the rise in R and G. We have always expected α to rise with frequency because of line capacitance, now we are in a position to calculate it.

The effect of β on wave velocity is brought out in (ii) of the Figure, ν changes noticeably at the lower frequencies, especially over the audio spectrum. At 100 Hz, ν = 15,700 miles/sec, at 10,000 Hz, 106,500 miles/sec. Consider a uniform line, say 100 miles in length transmitting music signals. At 100 Hz the propagation time t is

$$\frac{\text{distance}}{\text{velocity}} = \frac{100}{15{,}700} \text{ secs} = 6.37\text{ms.}$$

At 10,000 Hz, t is 0.94 ms and now we meet another type of distortion known appropriately as *delay/frequency distortion* because some frequencies are delayed more than others. Although a few ms is short in human experience, it is long compared with the periodic times of the waves being considered, accordingly over this circuit the received music signal would be of poor quality. Likewise a pulse transmitted over such a circuit would experience differing phase-shifts in its harmonic components, and hence arrive spread out in time.

Lastly at (iii) on Fig. 6.6 is illustrated the effect of frequency on the characteristic impedance Z_0. This is plotted in the form $R + jX$ and again by simply looking at the modulus of Z_0 at say, 2 and 10 kHz as shown dotted, there is another surprise for $|Z_0|$ changes from 333 to 170Ω, almost to half. So what has been said about matching the line to its termination seems to be an impossibility unless the frequency range is narrow. Fortunately the line characteristics can be modified so that all distortions are much reduced.

It can be shown mathematically that for minimum attenuation $LG = RC$ and for the particular line we are using as an example, at 1600 Hz

$$LG = 1.11 \times 10^{-3} \times 1.75 \times 10^{-6} \approx 1.9 \times 10^{-9}.$$

$$RC = 84.1 \times 0.0608 \times 10^{-6} \approx 5 \times 10^{-6}.$$

Evidently LG is much smaller than RC and this is generally so. RC cannot easily be reduced, hence LG must be increased, the most convenient for this being L. This is accomplished in practice by *loading* the cable with small inductors (*loading coils*, of values from 30 to 180 mH) connected in series with each wire at comparatively short intervals, 1000 or 2000 yards (0.91 or 1.83 km), an expensive procedure but one from which several advantages are gained. LG is still not made equal to RC but the difference is much less. In a way it looks at though the cable is being "tuned" by adding inductance so that the inductive reactance cancels the capacitive, but it is more complicated because of the presence of R and G.

Not only does this practice reduce the attenuation but theoretically:

(i) since $R + j\omega L = R\left(1 + \frac{j\omega L}{R}\right)$

and $G + j\omega C = G\left(1 + \frac{j\omega C}{G}\right)$

when $\frac{L}{R} = \frac{C}{G}$, $Z_0 = \sqrt{\frac{(R + j\omega L)}{(G + j\omega C)}} = \sqrt{\frac{R}{G}}$ or $\sqrt{\frac{L}{C}}$

neither of which has a j term nor includes ω, hence Z_0 is

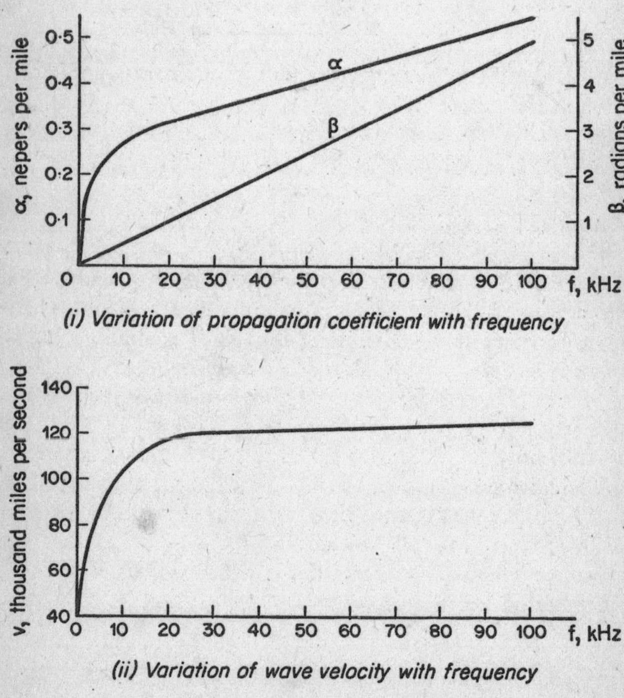

Fig. 6.6 Effect of frequency on line coefficients

resistive and independent of frequency.
(ii) with a little more manipulation, $\alpha = \sqrt{RG}$, $\beta = \omega\sqrt{LC}$ and

$$v = \frac{\omega}{\beta} = \frac{\omega}{\omega\sqrt{LC}} = \frac{1}{\sqrt{LC}}$$ and there is no frequency

term in any of these, hence α, β and v are all independent of frequency.

When LG = RC, this is known as the *distortionless condition* for the particular line.

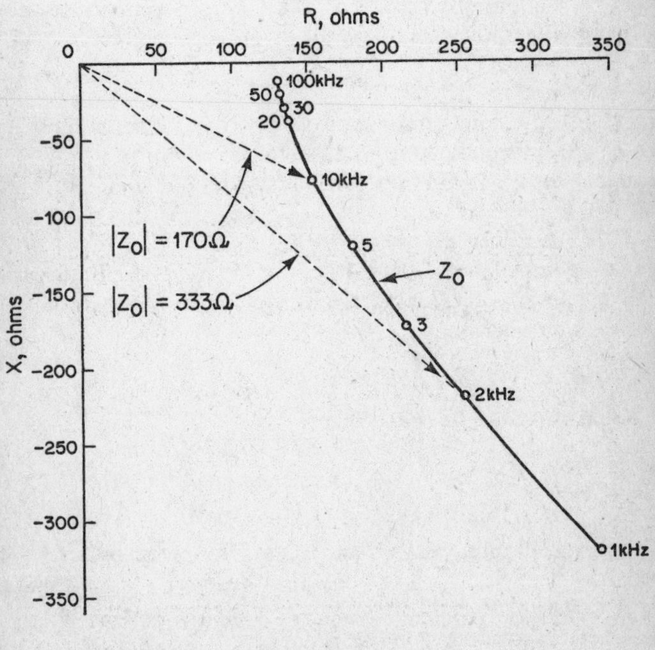

(iii) Variation of characteristic impedance with frequency

There is one more price to pay for these advantages apart from the cost of the loading coils. Because the loading is "lumped" even though at short intervals (theoretically it should be continuously distributed), the line behaves as a low-pass filter with an attenuation below that which it would have with no loading but only up to a certain frequency when the attenuation rises sharply, the "cut-off" point. This is advantageous for signals for which the highest frequency is below the cut-off point, but certainly not for multiplex systems, hence loading is used mainly for audio and music lines. Paradoxically telecommunications administrations are removing loading coils from existing cables so that p.c.m. systems can be used over them for the last thing such a system needs is a heavily restricted cable frequency response. The original loading points which need a manhole or underground chamber to accommodate the coils are now being re-employed to house the p.c.m. regenerators.

6.1.4 Reflection

Most of the previous considerations have been about lines theoretically infinitely long or the equivalent shorter line terminated in Z_0. In Section 6.1.1 arises the possibility of reflection, in the case mentioned the whole of the energy being reflected when the line meets an open-circuit, i.e. the terminating impedance $Z_t = \infty$. In general some reflection occurs whenever $Z_t \neq Z_0$, that is when a mismatch exists between the line and its termination.

Consider firstly a line *correctly terminated* as in Fig. 6.7(i) with a line current I_0 flowing, then

$$I_0 = \frac{Eg}{2Z_0} \quad \text{i.e. } Eg = 2I_0Z_0 \ldots\ldots\ldots \qquad \text{(i)}$$

Next consider the line to be incorrectly terminated by an impedance Z_t. I_0 is not completely absorbed by Z_t hence some current (I_r) is reflected giving rise to a total line current ($I_0 + I_r$). The artful move in this proof is now to consider Z_t to be split up into two impedances, the correct (Z_0) and the mismatch value Z_r such that

$$Z_t = (Z_0 + Z_r) \ldots\ldots \qquad \text{(ii)}$$

[See Fig. 6.7(ii)] so that we can find the voltage of a generator which would produce I_r. The Compensation Theorem (Section 3.1.2) comes to our aid in this for it shows that Z_r can be replaced by a generator of zero impedance and of voltage $-(I_0 + I_r)Z_r$ as shown to the right of Fig. 6.7(ii). The net voltage acting in the overall circuit is now $Eg - (I_0 + I_r)Z_r$ and the Ohm's Law equation becomes

$$(I_0 + I_r) = \frac{Eg - (I_0 + I_r)Z_r}{2Z_0}$$

$$\therefore\ 2Z_0(I_0 + I_r) = Eg - Z_r(I_0 + I_r)$$

$$\therefore\ Eg = (I_0 + I_r)(2Z_0 + Z_r)$$

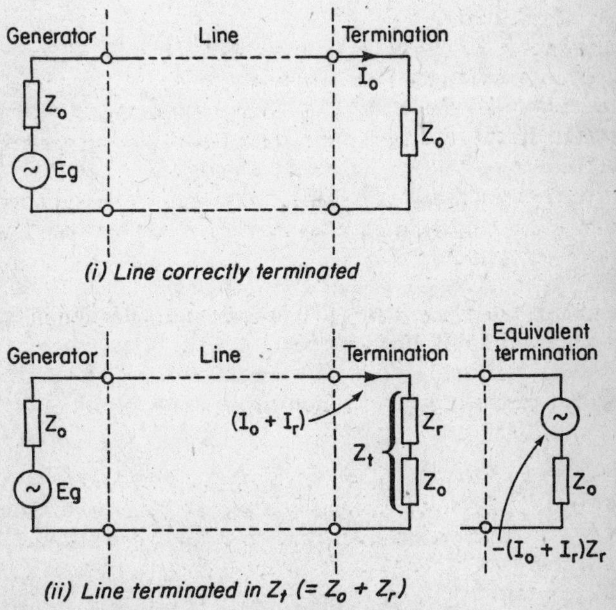

Fig. 6.7 Reflection on a transmission line

From equation (i)

$$2I_0Z_0 = 2I_0Z_0 + I_0Z_r + 2I_rZ_0 + I_rZ_r$$

$$\therefore -I_0Z_r = I_r(2Z_0 + Z_r)$$

$$\therefore \frac{I_r}{I_0} = \frac{-Z_r}{2Z_0 + Z_r}$$

and from equation (ii)

$$\frac{I_r}{I_0} = \frac{-(Z_t - Z_0)}{Z_0 + Z_t} = \frac{Z_0 - Z_t}{Z_0 + Z_t}$$

$\frac{I_r}{I_0}$ is the ratio of reflected to incident current and the expression

$\frac{Z_0 - Z_t}{Z_0 + Zt}$ is known as the *current reflexion coefficient.*

The return current I_r is therefore zero when $Z_t = Z_0$ (correctly terminated) for

$$I_r = I_0 \times \frac{Z_0 - Z_t}{Z_0 + Z_t} = 0$$

Also it is equal to $-I_0$ when $Z_t = \infty$ (open circuit) and $+ I_0$ when $Z_t = 0$ (short-circuit), that is complete reflection in both cases but with 180° phase change in the first case. When Z_t has other values I_r has a value between 0 and I_0. We look at a practical case in which reflection writes its own story in the next section.

6.2 COAXIAL TRANSMISSION LINES

There is little change in the basic transmission line theory when coaxial cables are considered even though the construction of the line is very different (see Section 1.6). Thus the coaxial cable has primary constants, R, L, G and C and the main formulae apply with no adjustments. In calculations however, the fact that at the higher frequencies, for which coaxial cables are particularly suited, $\omega L \gg R$ and $\omega C \gg G$ allows us to reduce the formula for Z_0 to $\sqrt{L/C}$ for air-dielectric cables. Z_0 would then appear to be independent of frequency since ω is not included in the formula. This in fact is true, it is R and G which vary most with frequency in any particular case but as mentioned, these are both swamped by L and C which vary very little. Taking a practical 6.8mm cable (inner diameter of outer conductor) as an example with L = 0.29 mH/km and C = 0.048μF/km at 100 kHz,

$$Z_0 = \sqrt{\frac{L}{C}} = \sqrt{\frac{0.29 \times 10^{-3}}{0.048 \times 10^{-6}}} = 78\Omega,$$

a result almost identical with that obtained when R and G are also taken into account. Generally Z_0 for coaxial cables lies between 50 and 100Ω.

Because the same formulae apply, coaxial cables need to be matched to the termination not only for maximum signal power to be delivered but also to minimize reflection. After one reflection from the termination a second will occur from the generator if this is also mismatched, the reflected current dying away as it experiences the attenuation on each traverse of the line.

It is instructive to examine a practical case in which reflection can be visibly troublesome. Consider a television receiver fed from an aerial via say, 20 metres of coaxial cable and assume that the matching is poor at both ends, thus reflection of the signal occurs. Let the velocity of the signal be 3×10^8 m/s (at these frequencies it is only a little less), then the time taken for the reflected current to reach the aerial and be reflected back again to the receiver:

$$= \frac{20 \times 2}{3 \times 10^8} \times 10^6 \mu s = 0.133 \mu s.$$

Now suppose the screen is 50cms wide so that when the spot

traverses one line it has travelled this distance. At say, 625 lines to one picture and 25 pictures per second, one line is set up in

$$\frac{10^6}{625 \times 25}\ \mu s = 64\mu s$$

$$\text{hence, spot velocity} = \frac{50\text{cms}}{64\mu s} = 0.78\text{cm}/\mu s$$

and a second image due to the reflection will appear at 0.78 x 0.133cms = 1.04mm behind the true image, so causing a "ghost" picture displaced by 1.04mm to the left of the main picture (the spot moves left to right). The ghost image is fainter because of the additional attenuation which the reflected signal suffers. This is not often experienced because both aerials and receivers are matched to standard coaxial t.v. feeders. Such a feeder might have an outside diameter of about 6mm.

Coaxial cable sizes run from some 3–4mm to 20 or more centimetres, for most purposes less than about 3cm. The attenuation, α, can be shown to be approximately proportional to $\sqrt{f}$ and inversely proportional to the cable dimensions, hence the reason for using large diameter cables when low attenuation is essential.

6.3 RADIO WAVES

We next become involved a little more deeply in the intricacies of changing electric and magnetic fields and of electromagnetic radiation. Our approach must be simple for the full analysis is most complex but at least we shall develop some appreciation of wave energy and how it leaves an aerial and travels into free space.

6.3.1 The Electromagnetic Field

Consider two straight in-line wires separated at the centre as shown in Fig. 6.8(i) and again in (ii) where a battery at the

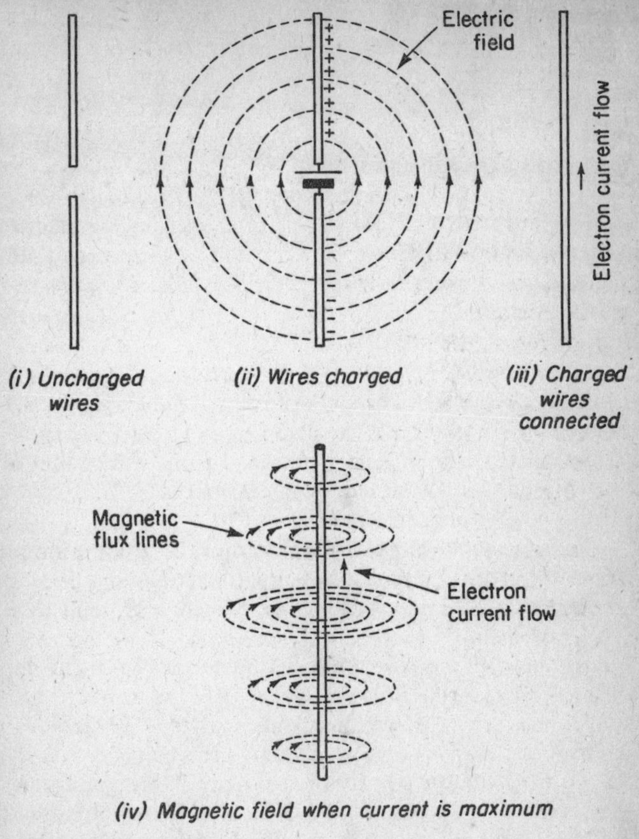

Fig. 6.8 Electric and magnetic fields

centre displaces electrons from the upper wire to the lower causing positive and negative charges as shown. The system is acting in fact as a simple capacitor. An electric field(1/4.2.1) exists between the two wires as shown, the arrows on the lines indicating the direction in which a free electron would be forced to move. Next suppose the battery to be removed and replaced by a short-circuit whereupon electrons immediately flow from the bottom to the top wire as in (iii). This current sets up a magnetic field(1/5.2.1) around the wire which is shown in (iv) and this field is at right-angles to the electric field. Because the charge on the wires is collapsing, so too is the electric field so giving way to the magnetic one. After reaching its maximum the wire current falls, the magnetic field collapses and creates a back-e.m.f.(1/5.3.1) which charges the wires in the opposite direction to that shown in (ii). The process so far is that of half a cycle of an oscillatory condition, the next half-cycle follows the same pattern but with field directions and current flow reversed. The fields soon die away because of losses.

The important features to recognize are that:

(i) the electric field alternates with the magnetic field
(ii) the two fields act at right angles
(iii) the magnetic field gives rise to the electric field but only while varying for only a changing magnetic field can produce the required e.m.f.
(iv) similarly a changing electric field must be present with the potential difference associated with setting up the magnetic field, that is, the two fields are interdependent.

Perhaps the most desirable picture for us to conjure up is the combination of Fig. 6.8(ii) and (iv) together so that we can visualize the two fields constantly changing over from one to the other and always at right angles. All this so far is over in a trice, but if the battery in (ii) of the figure is changed for an a.c. generator, then the fields will be maintained alternating at the frequency of the generator. It is not yet obvious how the wave is made to travel, this is the subject of the next section.

6.3.2 Radiation from an Aerial

Firstly we must dispel any notions that by merely increasing the voltage applied to the centre of the conductors in Fig.6.8 creates fields stretching out to such distances that radio communication is possible. It is actually fortunate that this does not happen otherwise all wires carrying a.c. voltages would radiate and radio communication would be in confusion.

6.3.2.1 *The Induction Field*

Consider the electric field and suppose there is a single tiny charge of Q coulombs (positive or negative, it does not matter which) out in the open. The electric flux is distributed in all directions outwards from the charge and at any distance r_1 from it can be considered as distributed over the surface of a sphere of radius r_1. Now the area of the surface of the sphere is $4\pi r_1^2$, hence the density of the flux at r_1 is $Q/4\pi r_1^2$ and further away at r_2 the same amount of flux is distributed over a surface $4\pi r_2^2$ with a flux density of $Q/4\pi r_2^2$ hence

$$\frac{\text{flux density at distance } r_2}{\text{flux density at distance } r_1} = \frac{\dfrac{Q}{4\pi r_2^2}}{\dfrac{Q}{4\pi r_1^2}} = \frac{r_1^2}{r_2^2}$$

As an example the flux density at a distance of 1m (r_2) from the charge relative to that at 1cm (r_1) is not 1/100 but $1^2/100^2$ = 0.0001, a very small value indeed. The same reasoning applies to any "point" source in space so this is known generally as the *inverse square law*, in this case simply saying that the flux density varies inversely as the square of the distance. Thus at any distance at all from an aerial the electromagnetic field which we have so far considered is useless for communication purposes. It is known as the *induction field* as distinct from the *radiated field* which travels outwards and is not governed by the inverse square law as we see next.

6.3.2.2 *The Radiated Field*

Consider again the two types of field in Fig. 6.8(ii) and (iv). The flux at the greatest distance must have expanded outwards from the centre and equally when the potential which brought this about is removed the flux collapses inwards towards the centre. Now this takes time, infinitesimally small though it is and the time = r/c where r is the distance in metres and c is the speed of light (3×10^8 m/s). But suppose that before the time for complete collapse has elapsed the voltage applied at the centre reverses and begins to set up a new field. This grows outwards and in opposition to the collapsing one, the latter is therefore repelled in front of the oncoming new field and travels outwards. The second field itself is repelled by that from the next change in polarity, the same is also happening to the magnetic field, hence an electromagnetic wave is projected outwards at the speed chosen by Nature long ago (note that light itself is an electromagnetic wave).

There is therefore quite a difference between the induction and radiated fields. The former returns its energy to the supply whereas with the latter the supply is constantly pumping energy in to replace that radiated, in other words power is continually flowing, not just a limited amount spreading out.

It can therefore be shown that the "strength" of the radiated wave varies inversely as the distance not as the square as for the induction field. Thus at great distances the radiated wave is useful whereas the induction field is not. Nevertheless the generator or radio transmitter usually has to supply kilowatts of signal power to maintain satisfactory signal/noise ratios at long distances.

We can now expand the mental picture of Fig. 6.8(ii) and (iv) combined. At some distance from the source the circles are so large that their appearance is like that of linear graph paper, the vertical lines representing the electric flux, the horizontal the magnetic flux, all lines appearing and disappearing according to the wave frequency and our mental arrows on them changing over accordingly. We are then looking at the *wavefront*. To go one step further we recall that the lines are our own naive way of representing a field, a mist gives a more realistic picture but this defies illustration on the printed page.

For Fig. 6.8 the wave is said to be *vertically polarized* because the *electric* field is vertical. If, however, on observing the wavefront we were to see horizontal electric flux lines, the wave would be *horizontally polarized*, such a wave is created by placing the two wires of Fig. 6.8(i) horizontally.

6.3.2.3 Dipole Aerials

*When the batt*ery is removed from Fig. 6.8(ii) and replaced by a *feeder* (a h.f. transmission line) connected to a generator the two wires then form a *dipole radiator* (aerial or antenna). Fig. 6.9 shows such an arrangement at the left and while the generator has the polarity shown electrons are extracted from the upper element and driven into the lower thus charges accelerate downwards in both elements and the dipole acts as a single wire. When the generator polarity reverses the charges accelerate upwards, overall giving the condition for radiation as we saw in Fig. 6.8. It follows then that a similar vertical dipole erected elsewhere will have its electrons displaced up and down by the effect of the electric field and a replica of the original signal is fed to the radio receiver. The figure shows one electrical direction of the field only and radiation in a single direction from the wires, also for clarity does not include the magnetic component. The latter can equally deliver the signal if its lines cut a loop of wire where the plane of the loop is at right angles to the flux lines.(1/5.3) From Fig. 6.9 it is also evident that if the receiving dipole is horizontal the induced e.m.f. is theoreticaly zero, that is a vertically polarized signal needs a vertical receiving aerial for maximum reception, if at an angle of less than 90° to the horizontal, then the induced e.m.f. is proportionately less.

The radio wave field strength is measured at any point in terms of the strength of the electric field per metre so if the receiving dipole of Fig. 6.9 were 1 metre long, the field strength figure would indicate the r.m.s. e.m.f. induced. At or near the transmitter we might talk in terms of volts per metre but at some distance away the strength would be quoted in mV or μV per metre.

Dipoles are not the only aerial configurations, there are many especially when directivity is required. We have chosen the dipole for study because it is an elementary (but very much used) aerial and one through which the fundamentals are most easily revealed.

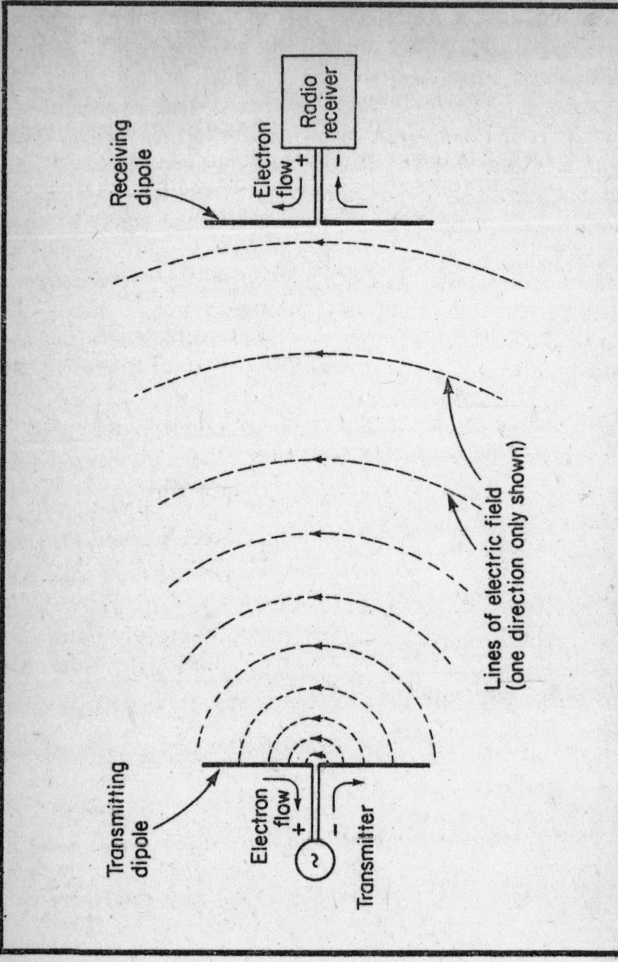

Fig. 6.9 Radio transmission using dipole aerials

6.4 WAVEGUIDES

Instead of transmitting electromagnetic waves into free space either on a broadcast basis or "point-to-point" using directional aerials, the wave can be confined within a metal guide of rectangular or circular cross-section. This has the advantage of a stable transmission medium with privacy but offset by cost of the guide and higher attenuation than for the radio wave. A sketch of a section of a rectangular type is shown in Fig. 6.10(i). The flange with holes is necessary for bolting sections together and the dimensions must be accurately maintained. Metals used are generally brass or aluminium.

The use of waveguides is restricted to the microwave band since the cross-sectional dimensions are related to the wavelength (at 10 GHz for example, $\lambda = 3$ cm). They are used both as feeders to and from microwave aerials (no energy should be

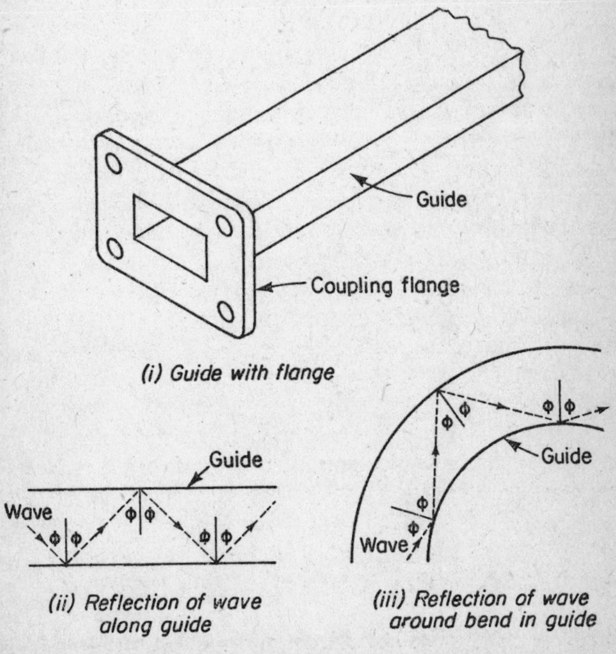

Fig. 6.10 Typical waveguide section

radiated except by the aerial) and also in microwave telephony systems where the multiplex signal is carried within tubular guides in the ground rather than suffer the hazards of the atmosphere. Such a tube is about 5cm in diameter and operates between about 40 and 100 GHz. A single tube can carry some 250,000 telephony channels or with a digital system, up to ten thousand megabits per second (10^{10}b/s).

Various methods are used to "launch" the wave into, or collect it from the guide, for example by extending the centre conductor of a coaxial cable carrying the signal into one end of the guide or by small rods acting in fact as tiny aerials.

It may be difficult to believe after having studied coaxial cables that transmission is entirely through the dielectric (usually dry air) and not along the walls of the guide but we must not lose sight of the fact that in this chapter we are studying electromagnetic fields and not current flow as in a transmission line. The normal radio wave which has already been considered cannot propagate straight along a guide because its electric field would be short-circuited wherever it ran parallel to the guide walls for these having very low resistance cannot have a significant potential difference anywhere (unless a very large current flows). Therefore various *modes* of propagation are used to avoid the electric field running parallel to the walls, these require the wave to travel in a zig-zag fashion along the guide, being reflected from wall to opposite wall (see Appendix 2, Section A2.1) on its journey as shown in Fig. 6.10(ii). The journey is therefore longer than if it were direct hence the velocity of propagation *along* the guide is less than that in free space. The method of propagation also allows negotiation of bends in the guide as shown in Fig. 6.10(iii).

7. OPTICAL TRANSMISSION

This is one of the most recent developments in communication technology and it has great promise. Without doubt the full theory is highly complex, nevertheless having now studied both multiplex systems and the electromagnetic wave there is no reason why we should not be able to get to grips with many of the system features. It is effectively an extension of wave-guide transmission but because of its future importance and the fact that it is part of the rather new science of *opto-electronics* it warrants a chapter on its own. For long-distance transmission the present optical fibre has such a value of attenuation that, for example, an undersea cable needs only a fraction of the number of amplifiers required by the normal coaxial system. There is nothing mysterious about optical transmission, it is simply that in the drive for higher and higher frequencies to accommodate wider bandwidths we have now jumped several orders of frequency from the E.H.F. band (around 10^{11} Hz) to the visible spectrum (see also Fig. 4.9) into a range extending approximately from 3×10^{14} to 8×10^{14} Hz (1,000 to 375 nm). Electromagnetic waves of these frequencies are capable of releasing electric charges within the human eye which the brain translates as light, the colour being determined by the actual frequencies within this band.

Optical transmission is accomplished (i) directly through the atmosphere, usually for specialized purposes (e.g. military) where it suffers many limitations as we well know when we ourselves are confronted by mist or fog or (ii) by glass or plastic fibre with little hindrance except through attenuation, it is this latter method which we examine because of its importance to multiplexed telephony, television and data signals.

7.1 THE OPTICAL FIBRE

This is usually a very fine circular strand of glass, less than 100μm in diameter, that is, somewhere around human hair thickness. It is purified to an extremely high degree to avoid light scattering. It is moderately flexible and many strands may be assembled within a small diameter plastic covered cable. The most easily understood type has a *cladding* surrounding it usually of a different high-quality glass with a slightly lower refractive index. This leads to *total internal reflection* as explained in detail in Appendix 2 (Section A2.2.1) and the ray propagates as for the waveguide of Section 6.4, this is pictured in Fig. 7.1. The transmission is termed *multi-mode* and it is evident that the separate electromagnetic waves will have slightly differing waveguide velocities so that when pulses are transmitted they could suffer some spreading. As with waveguides, bends in the fibre transmission line are negotiated without difficulty as depicted in Fig. 6.10(iii).

Such fibres have a loss of less than 4dB/km.

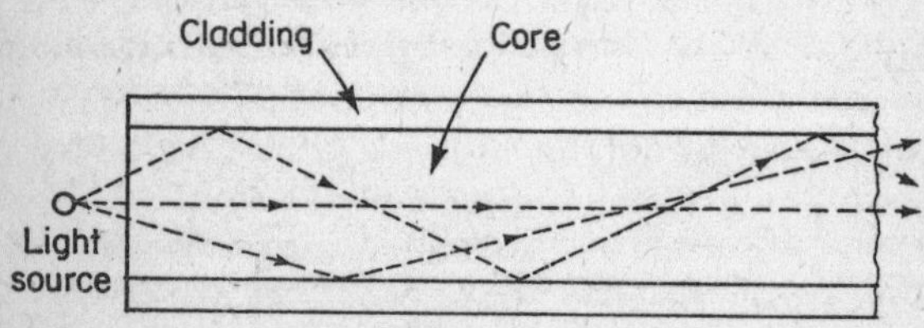

Fig. 7.1 Multimode transmission in fibre

7.2 LIGHT SOURCES

Of the many available, the most effective are lasers and electroluminescent diodes. To understand the operation of lasers requires a previous study of quantum electronics, much beyond us here, so we look at a typical type of laser in outline only for appreciation of just one of the light sources. LASER stands for "Light Amplification by Stimulated Emission of Radiation". The radiation is known as *coherent* (sticking together) for, as opposed to the normally seen non-coherent light which is a jumble of many different waves of varying amplitudes and phases, its components are all in phase and restricted to a single frequency or very small band. Stimulated emission implies firstly raising electrons to a high level of energy ("exciting" them) and then causing them to fall suddenly to a lower energy level so that each one ejects its surplus energy as a *photon* (a tiny particle of light). This happening to many electrons at once produces a very short duration, high energy, coherent light output. Exciting the electrons can be done by flooding them with intense light (such as from a photographic flash unit) when a transparent crystal is used, or by an electric current in the semiconductor laser, this is known as *pumping*.

The basic features of a semiconductor laser are shown in Fig. 7.2. Such a device operates most efficiently between about 600 and 1000 nm although they are quite effective well outside of this range. The very small size can be seen from the typical dimensions shown on the sketch. A pn junction is formed with contacts on the upper and lower surfaces of the block. The back surface is coated with a metallic reflector so that all radiation emerges from the junction at the front. The device is pumped by supplying electrons to the n-material so that they are forced across the junction. Recombination between electrons and holes having different energies produces the laser action and coherent radiation emerges from the thin junction line on the front face as shown.

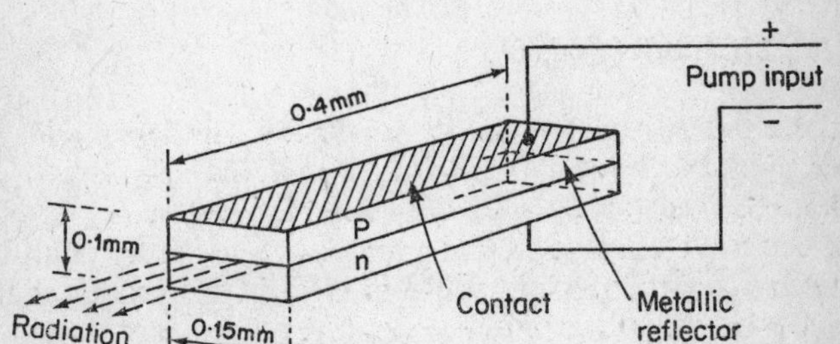

Fig. 7.2 Semiconductor laser

Modulation can take many forms according to the input signal, as a single example it is possible to modulate the biassing current to this type of laser. With optical pumping, its amplitude or frequency can be varied.

7.3 OPTICAL DETECTORS

As in the case of light sources, many different types exist, so we choose one which is not only easily recognized but has favourable characteristics for this application, a semiconductor junction detector, the *photodiode*. These are used in two different modes, *photovoltaic* in which incident light generates a voltage and *photoconductive* in which the device varies its resistance according to the incident light wavelength and intensity. The latter mode is preferred because it has a faster response to pulsed light (most systems are digital) and it also has a higher sensitivity.

The photodiode is a specially constructed semiconductor diode so that light can be effective on the junction and is reverse-biassed. With no illumination normal reverse-bias conditions obtain(3/1.5.1) so that a very small reverse saturation current flows (the *dark* current). With incident light, electrons absorb energy from it and conduction is increased, this is detected as a change of current through the device. This *photocurrent* rises linearly with the intensity of the illumination. Improved sensitivity is obtained by the avalanche effect(3/2.1.3) with the diode biassed just before the avalanche point. Electron-hole pairs created by photons are then able to create minor avalanche conditions to that there is a large increase of carriers flowing across the junction, hence the resistance change dark-to-light is magnified.

7.4 THE OPTICAL TRANSMISSION SYSTEM

From the above it is evident that coherent light sources and detectors available for use in optical transmission systems must be properly coupled to the fibre itself, an intricate task because of its extremely small diameter. The basic arrangement of one direction on a single optical fibre is given in Fig. 7.3. En route repeaters are needed to offset the fibre attenuation in the same way as applies for other multiplex systems, such a repeater

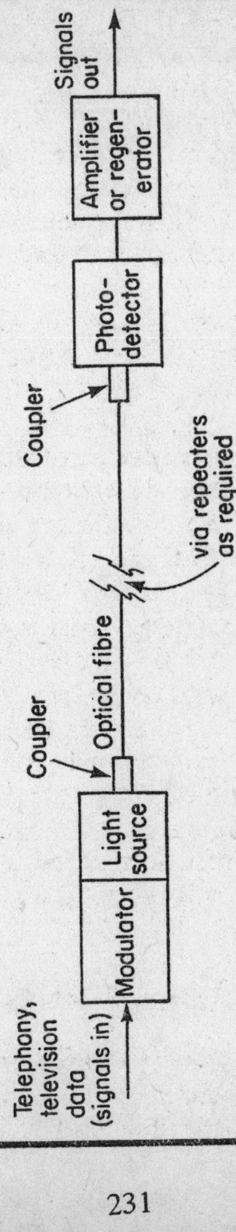

Fig. 7.3 Optical transmission system

comprises a photo-detector, regenerator (of the broadband signal), modulator and on-going light source. Repeater spacings are some 8 km or more and it is reasonable to predict digital system capacities some ten times that for waveguides, that is, one hundred thousand megabits per second (10^{11} b/s). In view of the convenience and lower cost of the fibre compared with that of a comparatively large metal tube, waveguides are likely to be superseded by fibres for long distance transmission.

On submarine systems more than 2000 circuits per pair of fibres appear possible, with repeater spacings of over 50 km.

APPENDIX 1

CONVERSION BETWEEN FREQUENCY AND WAVELENGTH

1	2	1	2	1	2	1	2	1	2	1	2
31	96.77	51	58.82	71	42.25	102	29.41	142	21.13	230	13.04
32	93.75	52	57.69	72	41.67	104	28.85	144	20.83	235	12.77
33	90.91	53	56.60	73	41.10	106	28.30	146	20.55	240	12.50
34	88.24	54	55.56	74	40.54	108	27.78	148	20.27	245	12.24
35	85.71	55	54.55	75	40.00	110	27.27	150	20.00	250	12.00
36	83.33	56	53.57	76	39.47	112	26.79	155	19.35	255	11.76
37	81.08	57	52.63	77	38.96	114	26.32	160	18.75	260	11.54
38	78.95	58	51.72	78	38.46	116	25.86	165	18.18	265	11.32
39	76.92	59	50.85	79	37.97	118	25.42	170	17.65	270	11.11
40	75.00	60	50.00	80	37.50	120	25.00	175	17.14	275	10.91
41	73.17	61	49.18	82	36.59	122	24.59	180	16.67	280	10.71
42	71.43	62	48.39	84	35.71	124	24.19	185	16.22	285	10.53
43	69.77	63	47.62	86	34.88	126	23.81	190	15.79	290	10.34
44	68.18	64	46.88	88	34.09	128	23.44	195	15.38	295	10.17
45	66.67	65	46.15	90	33.33	130	23.08	200	15.00	300	10.00
46	65.22	66	45.45	92	32.61	132	22.73	205	14.63		
47	63.83	67	44.78	94	31.91	134	22.39	210	14.29		
48	62.50	68	44.12	96	31.25	136	22.06	215	13.95		
49	61.22	69	43.48	98	30.61	138	21.74	220	13.64		
50	60.00	70	42.86	100	30.00	140	21.43	225	13.33		

$f \rightarrow \lambda$

Find the nearest most significant figures of f in Column 1	Multiply figure in Column 2 by	to give λ in
31 – 300 Hz	100	km
310 Hz – 3 kHz	10	km
3.1 – 30 kHz	1	km
31 – 300 kHz	100	m
310 kHz – 3 MHz	10	m
3.1 – 30 MHz	1	m
31 – 300 MHz	10	cm
310 MHz – 3 GHz	1	cm
3.1 – 30 GHz	1	mm

$\lambda \rightarrow f$

Find the nearest most significant figures of λ in Column 1	Multiply figure in Column 2 by	to give f in
3.1 mm – 3 cm	1	GHz
3.1 – 30 cm	100	MHz
31 cm – 3 m	10	MHz
3.1 – 30 m	1	MHz
31 – 300 m	100	kHz
310 m – 3 km	10	kHz
3.1 – 30 km	1	kHz
31 – 300 km	100	Hz
310 – 3000 km	10	Hz

Examples: 330m Most significant figures in Col.1 are 33. Adjacent figure in Col. 2 = 90.91. From $\lambda \rightarrow f$ table multiply by 10 to give 909.1 kHz.

960MHz. Most significant figures in Col. 1 are 96. Adjacent figure in Col. 2 = 31.25. From $f \rightarrow \lambda$ table multiply by 1 to give 31.25 cm.

APPENDIX 2

THE GEOMETRY OF ELECTROMAGNETIC RAYS

In this appendix we enquire into some of the things which happen to light rays in everyday life because they are relevant to certain sections in the main text. We look at

(i) *reflection* as from a mirror

(ii) *refraction* which makes a stick partly immersed in water appear bent and fish to seem nearer to the surface than they are

(iii) *total internal reflection*, when in a shop window we see an image of the street behind as if the glass were mirrored

(iv) the *parabola* which gives us the concentrated beam from headlamps, searchlights and pocket torches.

On the simple laws on which the above phenomena are based, directional microwave aerials (Secton 4.2.3.2), and optical fibre transmission (Section 7.1) are derived.

Firstly a reminder that whereas we talk of light *rays* and draw neat lines to represent them, this is merely our uncomplicated way of looking at things, it is seldom quite as simple as that.

A2.1 REFLECTION

In considering the paths of rays, angles are usually expressed at a point relative to the *normal* rather than to the surface which the ray strikes for this may not always be flat. The normal is simply a straight line drawn from the point of interest at right-angles to the surface as can be seen in Fig. A2.1.

The laws of reflection are:

(i) the incident, normal and reflected rays lie in the same plane, the two rays being on opposite sides of the normal

(ii) the angle of reflection is equal to the angle of incidence, both angles being relative to the normal and marked θ_i and θ_r in Fig. A2.1.

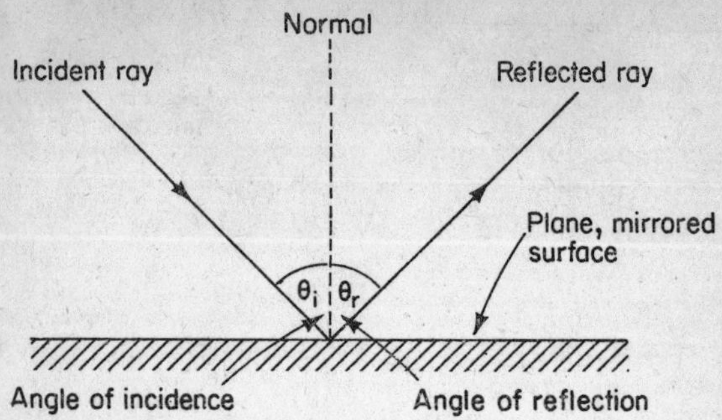

Fig. A2.1 Reflection

Light and radio waves are both electromagnetic waveforms (only differing in frequency) hence radio waves observe the same laws of reflection at suitable conducting surfaces.

A2.2 REFRACTION

In the foregoing section the ray of light travels in one medium only, usually air. When travelling in a medium other than a vacuum or air the velocity ν of the wave decreases to

$$\nu = \frac{c}{\sqrt{k}}$$

where $c = 3 \times 10^8$ m/s and k is the dielectric constant of the medium.(1/4.2.3)

We need not be concerned with actual values, what is important to us is that the velocity of light (or any electromagnetic wave) falls on entering a more dense medium.

Consider next a wavefront AB containing several rays as shown in Fig. A2.2. The incident rays 1–4 are arriving at the plane

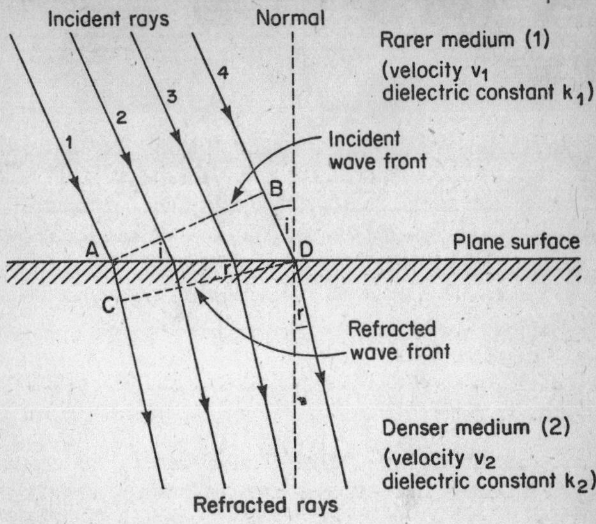

Fig. A2.2 Refraction

surface of a denser medium. The incident wavefront is shown at the instant when ray 1 is about to enter the denser medium and it is evident that rays 2, 3 and 4 have progressively further to travel in the rarer medium before they too enter the more dense. Thus between this moment and the time when ray 4 is at the surface, ray 1 in travelling more slowly has traversed the distance AC whereas ray 4 has moved through the greater distance BD. The *refracted* wavefront CD is shown dotted and it is clear that the whole wave is bent so that the angle of incidence i (as shown for ray 4) is greater than the angle of refraction r. On entering a more dense medium therefore, a ray is bent towards the normal. Now in Fig. A2.2, $\angle BAD = i$ and $\angle ADC = r$

$$\therefore \frac{\sin i}{\sin r} = \frac{\frac{BD}{AD}}{\frac{AC}{AD}} = \frac{BD}{AC}$$

and since BD and AC are proportional to the ray velocities ν_1 and ν_2 in the two mediums,

$$\frac{\sin i}{\sin r} = \frac{\nu_1}{\nu_2}$$

W. Snell (a Dutch astronomer and mathematician) discovered in the early 1600's that the ratio sin i/sin r is constant in any particular case for all values of i, and he called it the *index of refraction* (or *refractive index*). This is therefore defined for a substance as the ratio of the velocity of light in a vacuum (or in air, the velocity is almost the same) to its velocity in the substance. Using n for the index of refraction:

$$n = \frac{\nu_1}{\nu_2} = \frac{\sin i}{\sin r}$$

and it is of interest that, as shown above,

$$\nu_1 = \frac{c}{\sqrt{k_1}}, \quad \nu_2 = \frac{c}{\sqrt{k_2}}$$

$$\therefore \frac{\sin i}{\sin r} = \sqrt{\frac{k_2}{k_1}}$$

where k_1 and k_2 are the dielectric constants of mediums 1 and 2.

Two examples of n are 1.33 for water and about 1.5 for glass, this latter one is important to us in the study of optical fibre transmission. 1.5 is simply an average sort of value, there are many types of glass and as many refractive indices.

A2.2.1 Total Internal Reflection

Let us take the value of the index of refraction for glass as 1.5 as suggested above. Then for air to glass

$$\frac{\sin i}{\sin r} = n$$

and for glass to air

$$\frac{\sin i}{\sin r} = \frac{1}{n}$$

and we can examine the shift in the refracted ray as the angle of the incident ray increases, calculated from sin r = n.sin i, i.e. $r = \sin^{-1}(n.\sin i)$.(2/A3.2) Three incident rays A, B and C becoming refracted rays A′, B′, C′, are shown in Fig. A2.3 at increasing angles. The refracted ray in each case is determined from the above formula. It is evident that at some value of i, r becomes 90° (as in the case of ray C) and the refracted ray travels along the glass surface. Thus for angles of i above this value a ray cannot appear in the air refracted, it is in fact reflected back at the surface as shown for example for the ray D and following the normal rules for ***reflection***. The edge of the glass then behaves as a perfect mirror to rays striking the surface at angles to the normal greater than that of ray C,

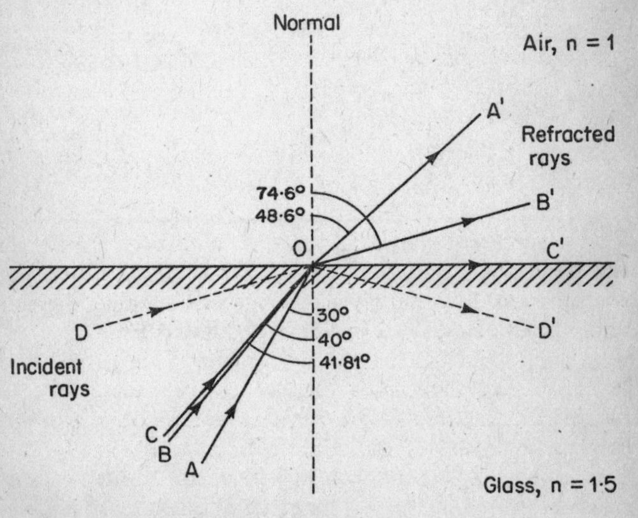

Fig. A2.3 Critical angle of refraction

which is said to be at the *critical angle*. This angle is easily calculated for when $r = 90^\circ$, $\sin r = 1$ and $\sin i = 1/n$ so in this case, for our glass, $\sin i = 1/1.5$

$$\therefore \; i = \sin^{-1} 0.6667 = 41.81^\circ$$

Total internal reflection is the principle used in prism binoculars and in many other optical devices. Its importance to us however is that we can now understand how a ray of light can travel along a tiny glass fibre without escaping from the sides. Fig. A2.4 shows how, θ_i must be greater than 41.81° for total internal reflection to occur (in glass of refractive index 1.5), θ_r is then equal to θ_i and the progress of the ray can be seen, there is theoretically no loss at each reflection point.

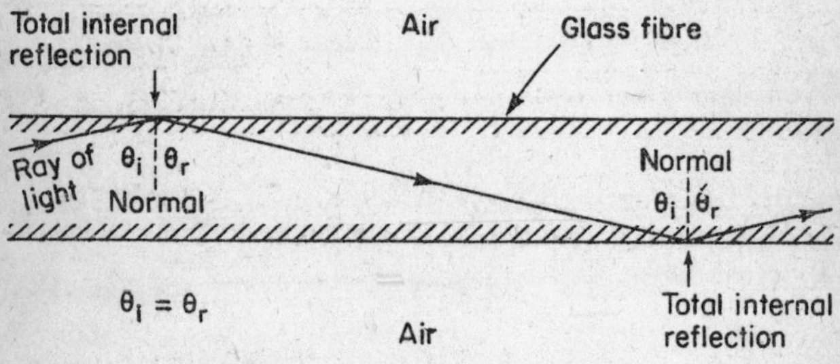

Fig. A2.4 Total internal reflection in glass fibre

A2.3 WAVE REFLECTION BY A PARABOLA

The use of a parabolic reflector in microwave telephony is considered in Section 4.2.3.2 of the main text, backed up by Fig. 4.13. Here we look at the geometrical proof that radio waves or light emanating from the focus are formed into a parallel beam and conversely that a parallel beam arriving on the axis of a parabolic reflector is directed towards the focus. Fig. 4.13(ii) has two special points labelled, the *focus* and the *vertex*. The focus is a point chosen when the reflector is designed and the vertex is that point where the curve meets the axis.

The equation to the curve of any parabola is $y^2 = 4ax$ where a is the distance from the focus to the vertex. We can therefore sketch our own parabola once having chosen a. Suppose we let a = 2 (the choice and whether in metres, centimetres etc does not matter here), then

$$y^2 = 4ax \quad \therefore \; y = \pm\sqrt{4ax}$$ (note that there are two values of y for each value of x).

A table for plotting the graph might be:

x	0	0.5	1	2	3	4	5	6	7	8
4ax	0	4	8	16	24	32	40	48	56	64
y (±)	0	2.0	2.83	4.0	4.90	5.66	6.32	6.93	7.48	8.0

resulting in a parabolic curve as in Fig. A2.5. Clearly the shape of any other curve depends on the value of a, but it is always a parabola.

Next we taken any point P on the curve and we wish to demonstrate that if the parabola is a cross-section of a reflector of light or microwaves, an incoming ray R parallel to the axis and meeting the parabola at P will be reflected to the focus F. Draw the tangent at P, also the normal, extending this to cut the axis at M. Join FP. By geometry which is not complicated but rather long and uninspiring it can be shown that FP = FM, hence $\angle$FMP = $\angle$FPM.

Now because PR and the axis are parallel, $\angle$MPR = $\angle$FMP $\therefore \angle$FPM = $\angle$MPR (all equal angles are marked θ in Fig. A2.5) which shows that the angle of incidence is equal to the angle of reflection for ray R and the reflected ray passes through the focus F. Equally a wave or ray generated at F in the direction of P is reflected along a path parallel to the axis. Note that the proof is general, the same considerations apply wherever we choose the point P.

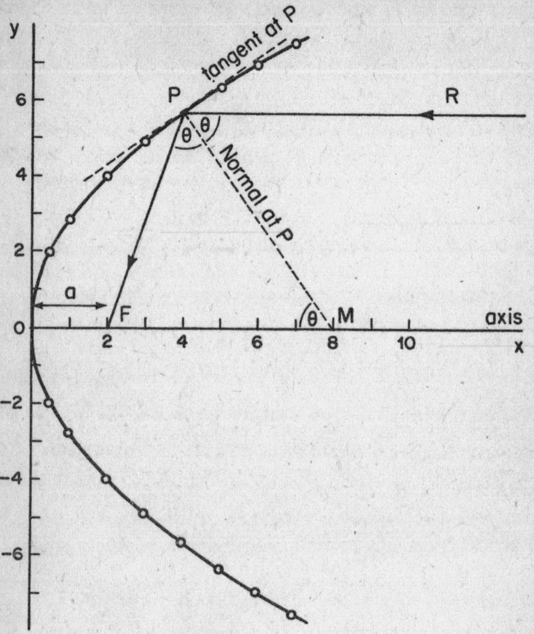

Fig. A2.5 Wave reflection by a parabola

APPENDIX 3

TRIGONOMETRY

For many trigonometry is a school subject, the purpose of which is always a mystery. Book 2 however in showing its relationship with the sine wave(2/A3) has perhaps brought some realization of its usefulness. Now we go just that little bit further to show how trigonometrical formulae help us to prove some of the communication principles for ourselves. Firstly however a little revision and then a simple method by which some of the more commonly met angles can be transformed into their trigonometrical functions without the need of tables.

A3.1 TRIGONOMETRICAL FUNCTIONS OF SOME KEY ANGLES

In the right-angled triangle, ratios between pairs of sides can be expressed as *trigonometrical functions* of the angles. If θ is one of the two angles other than the right-angle as in Fig. A3.1(i)

$$\sin\theta = \frac{\text{opposite side}}{\text{hypotenuse}} = \frac{BC}{AC},$$

$$\cos\theta = \frac{\text{adjacent side}}{\text{hypotenuse}} = \frac{AB}{AC}$$

$$\tan\theta = \frac{\text{opposite side}}{\text{adjacent side}} = \frac{BC}{AB}$$

where the hypotenuse is the side opposite the right-angle and is the longest. For the remaining $\angle ACB$ the adjacent side is BC and the opposite side AB. The values of these ratios are given in trigonometrical tables for angles from 0° to 90°. By inspection we can see that

$$\sin 0^\circ = 0 \qquad \sin 90^\circ = 1$$
$$\cos 0^\circ = 1 \qquad \cos 90^\circ = 0$$
$$\tan 0^\circ = 0 \qquad \tan 90^\circ = \infty$$

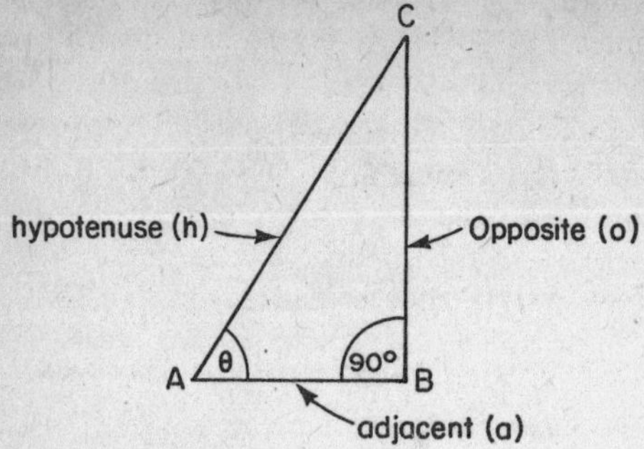

(i) Classification of sides relative to the angle θ

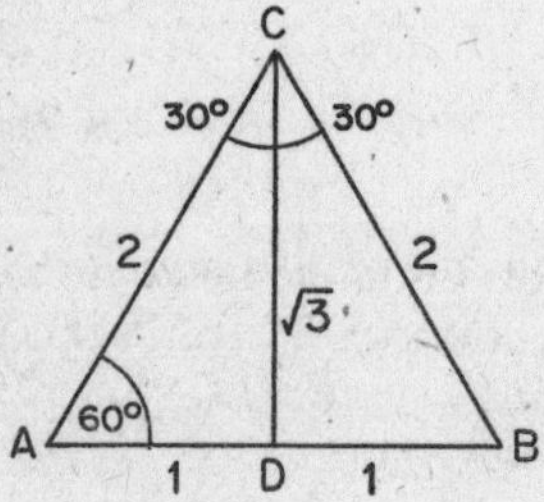

(ii) The 30/60 triangle

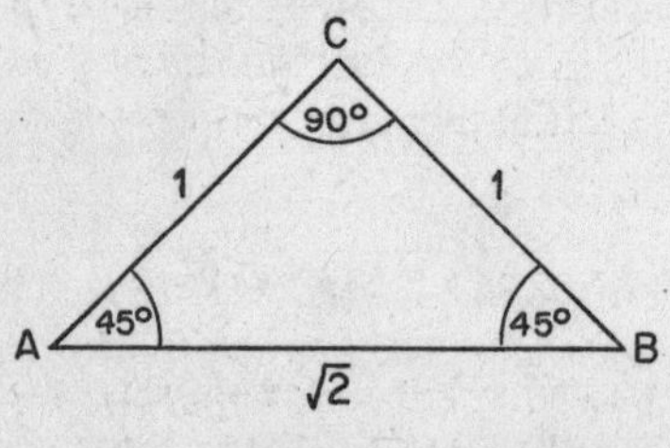

(iii) The 45/45 triangle

Fig. A3.1 Specimen triangles

It is also possible to recall from memory values for 30°, 45° and 60° by visualizing the two triangles shown in Fig. A3.1(ii) and (iii). At (ii) are shown what we might call two 30/60 triangles obtained by halving an equilateral one (all sides equal) and therefore equi-angular at 60° per angle. We conveniently choose to mark its sides as each of 2 units in length, hence by drawing the line perpendicular to AB from C we have:

$$AB = BC = AC = 2, \;\; AD = BD = 1 \;\; \angle ACD = \angle BCD = 30°$$

From Pythagoras' Theorem(2/A3.3) $AC^2 = AD^2 + CD^2$ from which $CD = \sqrt{3}$. Then

$$\sin 30° = \frac{1}{2} \qquad \cos 30° = \frac{\sqrt{3}}{2} \qquad \tan 30° = \frac{1}{\sqrt{3}}$$

$$\sin 60° = \frac{\sqrt{3}}{2} \qquad \cos 60° = \frac{1}{2} \qquad \tan 60° = \sqrt{3}$$

The 45/45 triangle of (iii) in the figure is an *isoceles* (two sides equal) triangle with one angle 90° as shown. If AC = BC = 1 then from Pythagoras' Theorem $AB = \sqrt{2}$. Then

$$\sin 45° = \frac{1}{\sqrt{2}}, \quad \cos 45° = \frac{1}{\sqrt{2}}, \quad \tan 45° = 1.$$

thus should we ever be uncertain of the shape of the sine, cosine or tangent curves, we have sufficient points on the way from 0 to 90° to sketch them for ourselves.

A3.2 MULTIPLE ANGLE FORMULAE

In Chapter 5 the need arises for expressing trigonometrical combinations in a different form. There are many formulae for this and here we develop some but mainly with a view to helping to prove our own work in the main text. We first recall why such formulae are necessary, a little thought shows that sin (A + B) is *not* the same as sin A + sin B nor is sin A.sin B equal to sin A.B, that is, the normal rules for solving mathematical equations cannot all be extended to trigonometry. Consider the two angles A and B shown in Fig. A3.2 in the right-angled triangle OPQ.

Then sin POQ, i.e. $\sin(A + B) = \dfrac{PQ}{OQ}$

Extend OR to S and draw QS so that $\angle RSQ$ is a right-angle

Draw a perpendicular ST from S onto PQ and SX onto OX.

Then $\angle OSX = (90^\circ - A)$ $\therefore$ $\angle RST = A$ (TS and OX are parallel)

Also $\angle TSQ = (90^\circ - A)$ $\therefore$ $\angle TQS = A$

Now

$$\sin(A + B) = \frac{PQ}{OQ} = \frac{PT + QT}{OQ} = \frac{PT}{OQ} + \frac{QT}{OQ}$$

also $PT = SX = OS \sin A$ and $QT = QS \cos A$

Hence

$$\sin(A + B) = \frac{OS}{OQ} \sin A + \frac{QS}{OQ} \cos A \quad \text{and since}$$

$$\frac{OS}{OQ} = \cos B \quad \text{and} \quad \frac{QS}{OQ} = \sin B$$

$$\sin(A + B) = \sin A.\cos B + \cos A.\sin B \ldots \qquad \text{(i)}$$

By using similar techniques it can be shown that:

$$\sin(A - B) = \sin A.\cos B - \cos A.\sin B \ldots \qquad \text{(ii)}$$

$$\cos(A + B) = \cos A.\cos B - \sin A.\sin B \ldots \qquad \text{(iii)}$$

$$\cos(A - B) = \cos A.\cos B + \sin A.\sin B \ldots \qquad \text{(iv)}$$

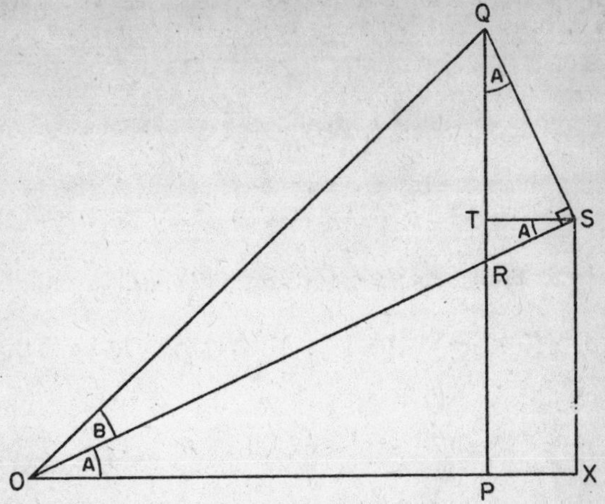

Fig. A3.2 Trigonometrical addition of two angles

From these relationships others follow, specifically:

subtracting equation (iii) from (iv) gives

$$2 \sin A.\sin B = \cos (A - B) - \cos (A + B)$$

$$\therefore \sin A.\sin B = \frac{1}{2} \cos (A - B) - \frac{1}{2} \cos (A + B) \ldots \qquad \text{(v)}$$

which is required to develop the amplitude modulation formula in Section 5.2.

Adding equations (i) and (ii)

$$2 \sin A.\cos B = \sin (A + B) + \sin(A - B)$$

and by substituting

$\frac{C + D}{2}$ for A and $\frac{C - D}{2}$ for B,

$A + B = C, \ A - B = D$

$$\therefore \ \sin C + \sin D = 2 \sin \frac{(C + D)}{2} \cdot \cos \frac{(C - D)}{2} \ \ldots \qquad \text{(vi)}$$

as required in Section 5.1 for the addition of two sine waves.

Generally therefore by recalling equations (i) to (iv) in the condensed form

$$\sin(A \pm B) = \sin A.\cos B \pm \cos A.\sin B$$

$$\cos(A \pm B) = \cos A.\cos B \mp \sin A.\sin B$$

we can then add or subtract pairs of these to obtain other relationships in which trigonometric forms of angles are either added or multiplied together. For example, by letting A = B,

$$\cos 2A = \cos^2 A - \sin^2 A \quad [\text{remember } \cos^2 A \text{ is shorthand for } (\cos A)^2]$$

Now from Fig A3.1(i) since

$$\sin^2\theta = \frac{o^2}{h^2} \text{ and } \cos^2\theta = \frac{a^2}{h^2}, \text{ then}$$

$$\sin^2\theta + \cos^2\theta = \frac{o^2 + a^2}{h^2}$$

which is equal to 1 because from Pythagoras' Theorem

$$o^2 + a^2 = h^2$$

We are calling the angle A instead of θ, hence $\cos^2 A = 1 - \sin^2 A$

$$\therefore \ \cos 2A = 1 - \sin^2 A - \sin^2 A = 1 - 2\sin^2 A$$

$$\therefore \ \sin^2 A = \frac{1}{2}(1 - \cos 2A) \ldots \qquad \text{(vii)}$$

which is needed for Sect. 5.2.3.1.